AF348969

Intelligent Systems Reference Library

Volume 161

Series Editors

Janusz Kacprzyk, Polish Academy of Sciences, Warsaw, Poland

Lakhmi C. Jain, Faculty of Engineering and Information Technology, Centre for
Artificial Intelligence, University of Technology, Sydney, NSW, Australia;
Faculty of Science, Technology and Mathematics, University of Canberra,
Canberra, ACT, Australia;
KES International, Shoreham-by-Sea, UK;
Liverpool Hope University, Liverpool, UK

The aim of this series is to publish a Reference Library, including novel advances and developments in all aspects of Intelligent Systems in an easily accessible and well structured form. The series includes reference works, handbooks, compendia, textbooks, well-structured monographs, dictionaries, and encyclopedias. It contains well integrated knowledge and current information in the field of Intelligent Systems. The series covers the theory, applications, and design methods of Intelligent Systems. Virtually all disciplines such as engineering, computer science, avionics, business, e-commerce, environment, healthcare, physics and life science are included. The list of topics spans all the areas of modern intelligent systems such as: Ambient intelligence, Computational intelligence, Social intelligence, Computational neuroscience, Artificial life, Virtual society, Cognitive systems, DNA and immunity-based systems, e-Learning and teaching, Human-centred computing and Machine ethics, Intelligent control, Intelligent data analysis, Knowledge-based paradigms, Knowledge management, Intelligent agents, Intelligent decision making, Intelligent network security, Interactive entertainment, Learning paradigms, Recommender systems, Robotics and Mechatronics including human-machine teaming, Self-organizing and adaptive systems, Soft computing including Neural systems, Fuzzy systems, Evolutionary computing and the Fusion of these paradigms, Perception and Vision, Web intelligence and Multimedia.

** Indexing: The books of this series are submitted to ISI Web of Science, SCOPUS, DBLP and Springerlink.

More information about this series at http://www.springer.com/series/8578

Ravi Patel · Dipankar Deb ·
Rajeeb Dey · Valentina E. Balas

Adaptive and Intelligent Control of Microbial Fuel Cells

 Springer

Ravi Patel
University of Auckland
Auckland, New Zealand

Rajeeb Dey
Department of Electrical Engineering
National Institute of Technology
Silchar, India

Dipankar Deb
Institute of Infrastructure Technology
Research and Management
Ahmedabad, Gujarat, India

Valentina E. Balas
"Aurel Vlaicu" University of Arad
Arad, Romania

ISSN 1868-4394   ISSN 1868-4408   (electronic)
Intelligent Systems Reference Library
ISBN 978-3-030-18067-6   ISBN 978-3-030-18068-3   (eBook)
https://doi.org/10.1007/978-3-030-18068-3

This Springer imprint is published by the registered company Springer Nature Switzerland AG
The registered company address is: Gewerbestrasse 11, 6330 Cham, Switzerland

Preface

Renewable and clean energy is a fundamental need of the human society today. At the same time, a third of the world population lacks adequate and cost-effective sanitation. Microbial Fuel Cells (MFCs) attempt to address both these needs by directly converting organic content to electricity from bacteria. It is now well known that electricity can be made using biodegradable material without even adding any special chemicals, simply by using bacteria already present in the wastewater with an anode compartment. The bacteria of a specifically designed fuel cell free of oxygen, attach to the anode. Because of lack of oxygen, they must transfer the electrons to the cathode rather than to oxygen. Then the electrons, oxygen, and protons combine to provide clean water. The electrodes, when at different potentials, create a fuel cell with influent food or "fuel" that is continuously used up by the bacteria.

This book provides basic information about fuel cells, and specifically MFCs, and the materials and the construction of such cells, all from the perspective of a control engineer in ensuring a regulated output from the cell. The book provides basic information about different modeling strategies applicable in MFCs, including outlines of statistical models, and more details with respect to engineering models. Mathematical models for single compartment MFCs with single microbial population and also dual microbial populations are described. Additionally, mathematical modeling for dual chamber MFCs is also presented. All these models are developed in a way that identifies the uncertain parameters and is appropriate for controller formulation such that the equivalent model is a control-oriented parametrized model appropriate for further control and estimation action.

The developed MFC model for single population is then individually analyzed for two types of inputs: (a) dilution rate and (b) influent concentration, with respect to the equilibrium points and the stability of those equilibrium points. A robust controller design for norm bounded uncertainty is studied for this single population MFC using Linearity Matrix Inequality (LMI) criterion.

Adaptive control methodologies are dealt with in detail in literature in the last half a century. However, in the context of MFCs, the authors felt the need to provide the basic formulations of the specific types of adaptive control methodologies with appropriate example that is relevant to the specific dynamical equations of the MFCs studied in this book. We also formulate, in detail, the dynamical equations representing single compartment MFCs containing single and dual populations, while systematically presenting the adaptive control methodologies most appropriate to the specific types of MFCs. An intelligent control method like the exact linearization method is presented for the more complicated MFC setup with dual chambers consisting of separate dynamical equations representing the anode and cathode chambers, but with a single population.

We have described a laboratory-level setup of five similar MFC setups with two-compartment configuration with cow dung slurry as the substrate. Using system identification techniques, the transfer function models of the cathode and anode compartments are developed, which is then controlled using two different MRAC techniques in the final chapter.

In summary, this book presents a systematic description of adaptive and intelligent control of different types of MFCs that span nonlinear control techniques like backstepping control, linear and robust control methods using linear matrix inequality, and estimation and adaptive update of uncertain parameters using adaptive control techniques. It is hoped that this book will facilitate a researcher to delve into the fundamentals of MFCs and position the researcher to develop and engineer the advanced control methods so as to able to make a well-presented handout for the formulation of new knowledge in this upcoming field of renewable energy.

The authors acknowledge the support received from Dr. Meenu Chabbra, Assistant Professor, Indian Institute of Technology Jodhpur and Dr. Sourav Das, Assistant Professor, Institute of Infrastructure Technology Research and Management, Ahmedabad, in developing the laboratory-level MFC setups which enabled validation of the control methodologies developed in this book.

Auckland, New Zealand   Ravi Patel
Ahmedabad, India   Dipankar Deb
Silchar, India   Rajeeb Dey
Arad, Romania   Valentina E. Balas

Contents

Contents   ix

About the Authors

Ravi Patel is currently a Research Scholar at the University of Auckland, New Zealand. He holds Master of Technology in Electrical Engineering from Institute of Infrastructure Technology, Research and Management, Ahmedabad, India. He holds B.E. degree from LDRP Institute of Technology & Research, Gandhinagar, India. He is a Student Member of IEEE and Asian Control Association. His research interest includes Adaptive Control, Nonlinear Control, and Renewable Energy and Microbial Fuel Cell and its control applications.

Dipankar Deb is a Professor and Department Coordinator in the Department of Electrical Engineering at Institute of Infrastructure Technology, Research and Management, Ahmedabad, India. He holds B.E. degree from National Institute of Technology, Karnataka, M.S. degree in Electrical and Computer Engineering from University of Florida, Gainesville, and Ph.D. degree in Electrical Engineering. He is a Senior Member of IEEE. He has filed and published 6 US patents and 25 Indian patents, apart from 50 research articles in International Journals and Conferences. He has authored and edited five books with Springer. He has wide experience both in Academia and Industry both in the US and in India. His research interest includes Adaptive Control, Active Flow Control, Renewable Energy, Smart Infrastructure, and Bio-medical Control Applications.

Rajeeb Dey is currently an Assistant Professor in the Department of Electrical Engineering at National Institute of Technology, Silchar, India. He holds M.Tech degree in Control System Engineering, from Indian Institute of Technology, Kharagpur, India and Ph.D. degree also in Control System Engineering from Jadavpur University, Kolkata, India. He is a Senior Member of IEEE Control System Society, Member Institution of Engineers (India), and Life Member of System Society of India. His research interest includes Design of Robust Control, Optimization based on LMI techniques, Time-Delay Systems, Intelligent Control, Decentralized Control and Control applications, and Bio-medical Control Applications.

Valentina E. Balas is currently a Professor in the Department of Automatics and Applied Software at the Faculty of Engineering, University "Aurel Vlaicu" Arad (Romania). She holds a Ph.D. in Applied Electronics and Telecommunications from Polytechnic University of Timisoara. She is author of more than 160 research papers in refereed journals and International Conferences. Her research interests are in Intelligent Systems, Fuzzy Control, Soft Computing, Smart Sensors, Information Fusion, Modeling, and Simulation. She is the Editor-in-Chief to International Journal of Advanced Intelligence Paradigms (IJAIP), member in Editorial Board member of several national and international journals, and is evaluator expert for national and international projects. Dr. Balas participated in many international conferences as General Chair, Organizer, Session Chair, and member in International Program Committee. She was a mentor for many student teams in Microsoft (Imagine Cup), Google, and IEEE competitions in the last years. She is a member of EUSFLAT, ACM and a Senior Member IEEE, member in TC—Fuzzy Systems (IEEE CIS), member in TC—Emergent Technologies (IEEE CIS), member in TC—Soft Computing (IEEE SMCS), and also a member in IFAC—TC 3.2 Computational Intelligence in Control. Dr. Balas Is Vice-president (Awards) of IFSA International Fuzzy Systems Association Council and Joint Secretary of the Governing Council of Forum for Interdisciplinary Mathematics (FIM)—A Multidisciplinary Academic Body, India.

Acronyms

1-D	One-Dimensional
2-D	Two-Dimensional
3-D	Three-Dimensional
CEM	Cation Exchange Membrane
DC	Dual Chamber
FC	Fuel Cell
LMI	Linear Matrix Inequality
MFC	Microbial Fuel Cell
MIT	Massachusetts Institute of Technology
MPC	Model Predictive Control
MRAC	Model Reference Adaptive Control
ODE	Ordinary Differential Equation
PD	Proportional & Derivative
PDE	Partial Differential Equation
PEM	Proton Exchange Membrane
PI	Proportional & Integral
PID	Proportional Integral & Derivative
SC	Single Chamber
SISO	Single Input Single Output
SPSC	Single Population Single Chamber

List of Figures

List of Tables

Chapter 1
Introduction

Rapid global industrial growth has exacerbated demand in energy supply among the increasing population. This demand is fulfilled from two main resources: (i) fossil fuels and nuclear energy [1], (ii) renewable energy sources. Energy derived from fossil fuels negatively impacts the environment by causing pollution and global warming. Such fuels may be unavailable in the near future, and so we must seek alternatives for the reduction of the dependency on non-renewable sources [2]. Renewable energy technologies have evolved over the years. These technologies interestingly are not dependent on the limited fuel sources. Energy extraction from organic or inorganic wastes efficiently resolves energy and environmental issues.

The large scale availability of energy to power devices is very important, and especially for remote or unattended applications, batteries or fuel cells are needed. The choice of power source for communication networks and instruments, depends on the environment and power needs of the device. For equipment placed in remote locations or intended for long term deployment, power sources like fuel cells that run for a long duration are advantageous.

Work on fuel cells began in the early part of the 19th century. While batteries release energy stored in a closed system, fuel cells are energy conversion systems that transfer electricity from replenishing sources of external fuel. Fuel cells may produce electricity continuously if provided with a sufficient flow of the external fuel, as opposed to batteries.

1.1 Fuel Cell

Fuel cells are static converters that combine features of engines and batteries to transform chemical energy into electrical energy, while providing an efficient and pure mechanism for energy conversions, with byproducts like water and heat [3]. Fuel cells have 80–85% efficiency while utilizing the generated heat. The operation

© Springer Nature Switzerland AG 2020

R. Patel et al., *Adaptive and Intelligent Control of Microbial Fuel Cells*, Intelligent Systems Reference Library 161, https://doi.org/10.1007/978-3-030-18068-3_1

Fig. 1.1 Basic construction
of FC

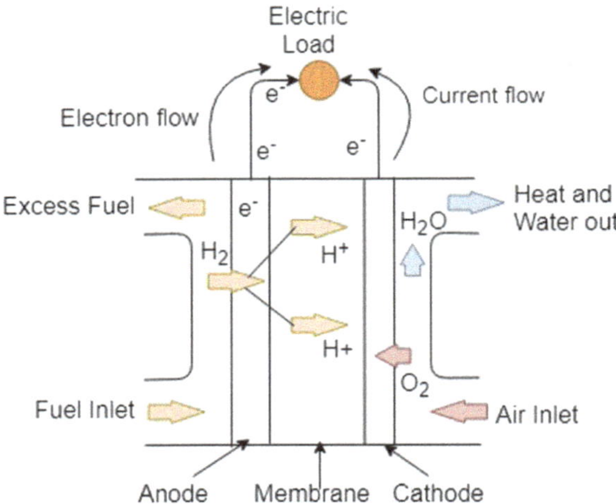

of fuel cells is noise and vibration free, with no moving part. Fuel cells have diverse
applications in stationary, portable and transportation power generation [4].

FCs consist of anode also referred to as fuel electrode, and cathode which is
called oxidant electrode, and electrolyte layer in contact with these two electrodes.
Hydrogen used as a fuel, is continuously fed to the anode and oxygen is supplied to
cathode. The hydrogen fuel gets oxidized at the anode electrode to produce positive
hydrogen ions and electrons. The electrolyte allows transmission of hydrogen ions,
and electrons are diverted to the external circuit and on to the cathode. The oxygen
from air, electrons and hydrogen ions are combined at cathode and produce water as
a byproduct [3].

The construction of a fuel cell with the basic features, is given in Fig. 1.1.

The chemical reactions of fuel cell are the following:

$$Anode\ Reaction:\ H_2 \rightarrow 2H^+ + 2e^-$$

$$Cathode\ Reaction:\ \frac{1}{2}O_2 + 2H^+ + 2e^- \rightarrow H_2O$$

$$Overall\ Reaction:\ H_2 + \frac{1}{2}O_2 \rightarrow H_2O$$

Fuel cells are classified as per the electrolyte type, fuel supply and types of membranes utilized [3]. Fuel cells have become emerging back up power sources for
telecommunication, material handling plants, airports and emergency services. Significant research work has been carried out on fuel cell technology advancement such
as hydrogen production, storage and transportation [3].

1.2 Microbial Fuel Cell

Microbial fuel cells (MFCs) or bio-fuel cells are promising pollutant removal units which use microorganisms as bio-catalytic elements for substrate oxidization and to produce electrical energy from chemical energy. MFC includes an anode compartment with an anode and a bio-catalyst and a cathode chamber with a cathode and another bio-catalyst. Typically, a proton exchange membrane (PEM) is positioned between the compartments to transfer protons, and an electrical pathway is provided between the two electrodes. The anode bio-catalyst catalyzes the oxidation of an organic substance, and the cathode bio-catalyst catalyzes the reduction of an inorganic substance. The reduced organic substance forms a precipitate, that removes the inorganic substance from solution. Bacterias in an anode chamber generate protons which are transferred to the anode so as to subsequently generate electrons using suitable substrates [10] and transfer these electrons via an external circuit. Direct transfer is a mechanism wherein electron flow takes place through mediator and biofilm at the anode [11]. In indirect electron transfer methodology, external mediator ensures electron transfer from the bacteria to surface of the anode [12]. At the cathode, protons and electrons combine to generate electricity, and produce fresh water as a byproduct.

Water pollution is severely increasing with industrial development, and so wastewater treatment has received increased attention over the years, but these are mostly inefficient, expensive and non-sustainable [5]. Wastewater composes of diverse microbial (fermentative, methanogenic and anodophilic) communities with an inherent capability to produce electrical energy. Such a process is more attractive than the wastewater treatment facilities because of higher efficiency of conversion and the capability to operate at ambient pressure and temperature, to provide bioelectricity [6–8]. MFCs are capable of cleaning wastewater with zero or positive energy [9], using inexpensive electrolytes, electrodes and urine, wastewater and other waste are used as fuel in anode chamber. MFCs are attractive choices for long-term power supply due to the longevity of the cells in a variety of applications for remote sensing and long duration studies.

1.3 Construction and Materials

Based on the need of the specific applications, MFCs are constructed in different ways. The performance as per electron transfer and electrochemical efficiency relies upon the electrode materials. For commercial use, the electrodes are required to be cost-effective, sustainable, readily available, and provide maximum power density at optimal cost. The biofilm formation and electron transfer from bacterias to anode surface primarily depends on the conductive and noncorrosive nature of anode materials. Simplest materials suitable for anodes are graphite plates which are inexpensive and also possess a well-defined surface area. A high degree of porosity is needed so

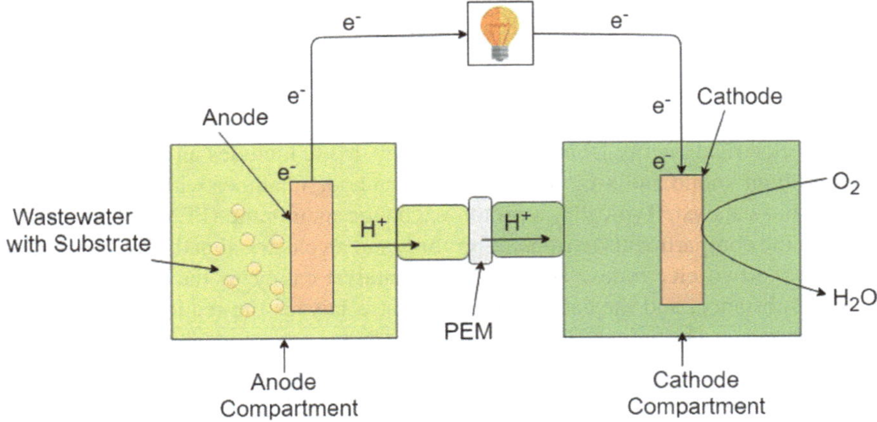

Fig. 1.2 Basic construction of two chamber MFC

that there is no clogging [13]. Carbon nanotubes are emerging as alternative for anode materials and possess good electrical conductivity and chemical stability. Surface and electrochemical oxidation treatments are possible over the anode surface to improve power density and bacterial adhesion [25]. The basic structure of a two-chamber MFC along with the associated components, is as given in Fig. 1.2.

The cathode material is a significant factor that decides system performance, and is chosen so as to operate near open circuit voltage potential. Ferricyanide is a usual candidate as a preferred electron acceptor [13], but it's diffusive properties affect long-term performance. Apart from ferricyanide, typical MFCs use hydrogen phosphate, oxygen, hydrogen peroxide, manganese dioxide, or copper chloride as the oxidizer. Typically, cathode used in a MFC contains three layers: diffusion layer, catalyst and conducting support material. Cathode material must have catalytic property, high mechanical strength and high electronic and ionic conductivity. Catalyst layer of platinum over carbon or graphite based cathode is utilized to improve the oxygen reduction reaction. Often, due exorbitant cost of platinum, the bacteria themselves act as catalyst called bio-cathode [25]. Different anode and cathode materials and the effects on system performance are analyzed and presented in [14–16].

The primary objective of MFCs is to remove pollutant from wastewater and generate electricity. The substrate is the most important biological factor which affect electricity generation. Performance is dependent on the concentration of substrate meant for oxidization, and type of substrate which are typically different types of wastewater containing microbes. Simple substrate is easy to degrade whereas composite substrates are difficult to degrade and help to develop electrochemically active microbial community in the system [10]. The performance of MFC using different types of substrates is analyzed and reviewed in literature [17–24].

The membrane physically separates anode and cathode compartments. Typically, Proton exchange membrane (PEM) facilitates flow of ions and also prevents penetration of substrate and oxygen. Nafion (class of synthetic polymer with ionic properties) is an often used PEM. Ion exchange membranes can impact system performance and stability of MFCs and so have seen a rapid growth in demand recently [13]. Advancement in membranes and their effect in performance of MFC are presented in [26–35]. Moreover, membrane-less MFCs are also in use with advantages such that the wastewater primarily treated in an anode compartment, can be used as an electrolyte in the cathode chamber. Therefore, a special electrolyte does not necessarily have to be utilized in the cathode compartment. In addition, the wastewater is aerobically treated in the cathode compartment. Generated electrons from bacterias are indirectly or directly transferred to the anode electrode. The yields obtainable from MFCs depend somewhat on the particular electron-carrier used. Direct transfer occurs through nanowires or intracellular mediator, and biofilm at the anode called "mediator-less MFCs" (bacteria as mediator) [11].

The usage of an external mediator that facilitates electron flow from bacteria to anode is known as the indirect method [12]. Performance enhancement of MFCs using different mediators are presented in [36–42]. Even with the advances made in MFCs, the process still produces only small electrical currents, far below the energy production of other fuel cells. What is needed is a more efficient electron transfer device. Current MFC technologies have major limitations such as moderate reliability, lower power density and high cost in terms of chemical process required to get optimal output from MFCs. A continuous improvement of materials and components is required for efficient performance. The development of advanced control strategies provides performance improvement in terms of efficiency, power density and consistent output voltage etc. MFCs are complex and nonlinear devices which require understanding of biological, electrochemical and thermal phenomena, so as to achieve closed-loop operation for efficient performance.

MFC parameters like temperature, pH value, pressure levels are well maintained and must be controlled to fulfill load requirements and ensure safe operation as well as avoid damage of parts. Performance of conventional linear control techniques may be affected by unmodeled dynamics, modelling error, different disturbances and parameter uncertainties and those controllers are needed for wide operation ranges. Reliable control technologies guaranteeing stability and robustness against uncertainty and external disturbance are of crucial importance for the advancement and practical usage of MFC technologies.

1.4 Scope and Outline of the Book

This book addresses the analysis and design of adaptive controllers for different MFCs. It advances the state-of-the-art and provides details for further research and development of MFC technologies. The content is conceptually divided in two parts. The first part is given to hypothesis and mathematical modelling of different MFCs. Basic operation, construction and components of MFC and the basic operation of conventional fuel cells are discussed. The mathematical modelling of MFCs in the literature is presented with special focus on suitability for control design.

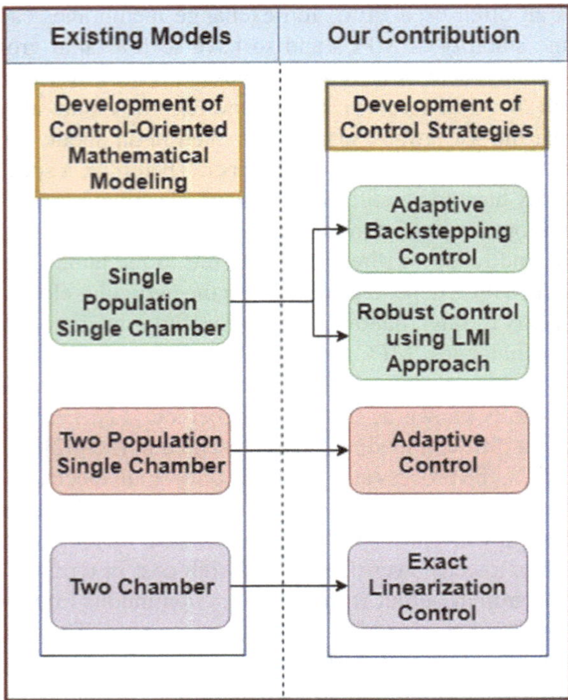

Fig. 1.3 Control development of existing MFC models

The linearization of single chamber MFC with single population model is given with equilibrium points and their stability through Jacobian matrices. A robust control technique is presented for SPSC MFC with norm bounded uncertainty using Linear matrix inequality (LMI) approach. The basics of adaptive controllers and in particular direct-indirect adaptive control, model reference adaptive control and adaptive backstepping control are summarized, and how such controllers are developed for the appropriate system are also discussed.

The second part presents an adaptive backstepping controller for a SPSC MFC and adaptive control of single chamber dual population MFC with simulation results. A detailed design procedure of adaptive controllers are outlined. Performance of both controllers are analyzed and discussed in depth. An exact linearization control technique is developed for two chamber MFC. A detailed procedure of making MFCs with materials in laboratory, is provided. Based on experimental data, transfer function model of MFC is obtained using system identification technique, and Model reference adaptive control (MRAC) techniques are developed.

Certain types of MFCs are not described in this book. For instance, soil-based MFCs containing soil as the nutrient-rich anodic media, and the proton exchange membrane (PEM), with the anode inserted at a certain depth within the soil, and the cathode is kept on the top and exposed to air [43]. Another type of MFC uses a

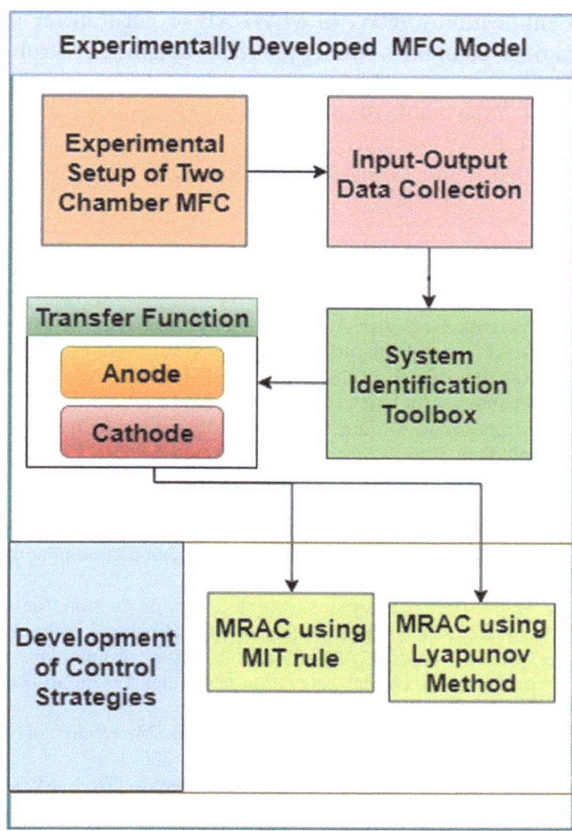

Fig. 1.4 Experimental development of MFC model and its control strategies

phototrophic biofilm anode containing photosynthetic microorganisms to undertake the process of photosynthesis, generate organic metabolites and donor electrons [44].

Generally, there are three types of MFCs available in literature such as single chamber MFC with single population and two population, and two chamber MFC. A schematic diagram of control techniques development of existing MFC models is shown in Fig. 1.3.

We have developed control-oriented mathematical model of three types of MFCs which provide state representation of models and their uncertain parameters. Suitable control techniques are developed as per performance requirement. An adaptive backstepping control and Robust control with LMI approach are developed for single population single chamber MFC. An adaptive control technique is applied to two population single chamber MFC. Exact Linearization control technique is developed for two chamber MFC.

Two chamber MFC is developed at laboratory level and input-output data is obtained. System identification is one of the technique to get mathematical models in different forms using experimental input-output data. We have used system identification toolbox in MATLAB to get transfer function models of anode and cathode chambers. Model reference adaptive control technique using MIT rule and Lyapunov method is developed separately for two chamber MFC. A schematic diagram of the whole development of MFCs model and its control strategies is given in Fig. 1.4.

References

1. Akdeniz, F., Çaglar, A., Güllü, D.: Recent energy investigations on fossil and alternative non-fossil resources in Turkey. Energy Convers. Manag. **43**, 575–589 (2002)
2. Rahimnejad, M., Adhami, A., Darvari, S., Zirepour, A., Oh, S.: Microbial fuel cell as new technology for bioelectricity generation: a review. Alex. Eng. J. **54**, 745–756 (2015)
3. Kirubakaran, A., Jain, S., Nema, R.: A review on fuel cell technologies and power electronic interface. Renew. Sustain. Energy Rev. **13**, 2430–2440 (2009)
4. Sharaf, O., Orhan, M.: An overview of fuel cell technology: fundamentals and applications. Renew. Sustain. Energy Rev. **32**, 810–853 (2014)
5. Hu, P., Ouyang, Y., Wu, L., Shen, L., Luo, Y., Christie, P.: Effects of water management on arsenic and cadmium speciation and accumulation in an upland rice cultivar. J. Environ. Sci. **27**, 225–231 (2015)
6. Mathuriya, A., Sharma, V.: Bioelectricity production from paper industry waste using a microbial fuel cell by Clostridium species. J. Biochem. Technol. **1**, 49–52 (2009)
7. Qin, M., Hynes, E., Abu-Reesh, I., He, Z.: Ammonium removal from synthetic wastewater promoted by current generation and water flux in an osmotic microbial fuel cell. J. Clean. Prod. **149**, 856–862 (2017)
8. Hao Yu, E., Cheng, S., Scott, K., Logan, B.: Microbial fuel cell performance with non-Pt cathode catalysts. J. Power Sour. **171**, 275–281 (2007)
9. Santoro, C., Arbizzani, C., Erable, B., Ieropoulos, I.: Microbial fuel cells: From fundamentals to applications: a review. J. Power Sour. **356**, 225–244 (2017)
10. Pant, D., Van Bogaert, G., Diels, L., Vanbroekhoven, K.: A review of the substrates used in microbial fuel cells (MFCs) for sustainable energy production. Bioresour. Technol. **101**, 1533–1543 (2010)
11. Lovley, D.: The microbe electric: conversion of organic matter to electricity. Curr. Opin. Biotechnol. **19**, 564–571 (2008)
12. Lovley, D.: Bug juice: harvesting electricity with microorganisms. Nature Rev. Microbiol. **4**, 497–508 (2006)
13. Logan, B., Hamelers, B., Rozendal, R., Schräuder, U., Keller, J., Freguia, S.: Microbial fuel cells: methodology and technology. Environ. Sci. Technol. **40**, 5181–5192 (2006)
14. Dumitru, A., Scott, K.: Anode materials for microbial fuel cells. Microb. Electrochem. Fuel Cells, 117–152 (2016)
15. Yamashita, T., Yokoyama, H.: Molybdenum anode: a novel electrode for enhanced power generation in microbial fuel cells, identified via extensive screening of metal electrodes. Biotechnol. Biofuels **11**(39), 1–13 (2018)
16. Mustakeem, M.: Electrode materials for microbial fuel cells: nanomaterial approach. Mater Renew. Sustain. Energy **4**(22), 1–11 (2015)
17. Gezginci, M., Uysal, Y.: The Effect of different substrate sources used in microbial fuel cells on microbial community. JSM Environ. Sci. Ecol. **4**(3) (2016)
18. Pandey, P., Shinde, V.N., Deopurkar, R.L., Kale, S.P., Patil, S.A., Pant, D.: Recent advances in the use of different substrates in microbial fuel cells toward wastewater treatment and simultaneous energy recovery. Appl. Energy **168**, 706–723 (2016)

19. Rezaei, F., Richard, T.L., Brennan, R.A., Logan, B.E.: Substrate-enhanced microbial fuel cells for improved remote power generation from sediment-based systems. Environ. Sci. Technol. **41**(11), 4053–4058 (2007)
20. Chae, K.J., Choi, M.J., Lee, J.W., Kim, K.Y., Kim, I.S.: Effect of different substrates on the performance, bacterial diversity, and bacterial viability in microbial fuel cells. Bioresour. Technol. **100**(14), 3518–3525 (2009)
21. Zhao, Y.G., Zhang, Y., She, Z., Shi, Y., Wang, M., Gao, M., Guo, L.: Effect of substrate conversion on performance of microbial fuel cells and anodic microbial communities. Environ. Eng. Sci. vpl. **34**(9), 666–674 (2017)
22. Wu, W., Yang, F., Liu, X., Bai, L.: Influence of substrate on electricity generation of Shewanella loihica PV-4 in microbial fuel cells. Microb. Cell Factories **13**(1), 1–6 (2014)
23. Mokhtarian, N., Rahimnejad, M., Najafpour, G.D., Daud, W.R.W., Ghoreyshi, A.A.: Effect of different substrate on performance of microbial fuel cell. African J. Biotechnol. **11**(14), 3363–3369 (2012)
24. Garba, N., Saadu, L., Balarabe, M.: An overview of the substrates used in microbial fuel cells. Greener J. BioChem. Biotechnol. **4**(2), 7–26 (2017)
25. Chouler, J., Bentley, I., Vaz, F., Fee, O., A, Cameron P, Di Lorenzo M.: Exploring the use of cost-effective membrane materials for Microbial Fuel Cell based sensors. Electrochimica Acta. **79**, 319–326 (2017)
26. Scott, K.: Membranes and separators for microbial fuel cells. Microb. Electrochem. Fuel Cells, 153–178 (2016)
27. Leon, J.X., Daud, W.R.W., Ghasemi, M., Liew, K.B., Ismail, M.: Ion exchange membranes as separators in microbial fuel cells for bioenergy conversion: a comprehensive review. Renew. Sustain. Energy Rev. **28**, 575–587 (2013)
28. Chouler, J., Bentley, I., Vaz, F., Fee, O., A, Cameron PJ, Di Lorenzo M.: Exploring the use of cost-effective membrane materials for Microbial Fuel Cell based sensors. Electrochimica Acta **231**, 319–326 (2017)
29. Das, S., Dutta, K., Rana, D.: Polymer electrolyte membranes for microbial fuel cells: a review. Polymer Rev., 1–20 (2018)
30. Ghassemi, Z., Slaughter, G.: Biological fuel cells and membranes. Membranes **7**(1), 1–12 (2017)
31. Rahimnejad, M., Bakeri, G., Najafpour, G., Ghasemi, M., Oh, S.: A review on the effect of proton exchange membranes in microbial fuel cells. Biofuel Res. J. **1**, 7–15 (2014)
32. Dharmadhikari, S., Ghosh, P., Ramachandran, M.: Synthesis of proton exchange membranes for dual-chambered microbial fuel cells. J. Serbian Chem. Soc. **83**(5), 611–623 (2018)
33. Zhang, X., Cheng, S., Huang, X., Logan, B.E.: Improved performance of single-chamber microbial fuel cells through control of membrane deformation. Biosens. Bioelectron. **25**, 1825–1828 (2010)
34. Christgen, B., Scott, K., Dolfing, J., Head, I.M., Curtis, T.P.: An Evaluation of the performance and economics of membranes and separators in single chamber microbial fuel cells treating domestic wastewater. PLOS ONE **10**(8) (2015)
35. Mishra, B., Awasthi, S., Rajak, R.: A review on electrical behavior of different substrates, electrodes and membranes in microbial fuel cell. World Academy Sci. Eng. Technol. Int. J. Energy Power Eng. **11**(9), 1023–1027 (2017)
36. Lohar, S., Patil, V., Patil, D.: Role of mediators in microbial fuel cell for generation of electricity and waste water treatment. Int. J. Chem. Sci. Appl. **6**(1), 6–11 (2015)
37. Park, D., Zeikus, J.G.: Electricity generation in microbial fuel cells using neutral red as an electronophore. Appl. Environ. Microbiol. **66**(4), 1292–1297 (2000)
38. Lin, C.W., Wu, C.H., Chiu, Y.H., Tsai, S.L.: Effects of different mediators on electricity generation and microbial structure of a toluene powered microbial fuel cell. Fuel **125**, 30–35 (2014)
39. Sund, C.J., McMasters, S., Crittenden, S.R.: Effect of electron mediators on current generation and fermentation in a microbial fuel cell. Appl. Microbiol. Biotechnol. **76**, 561–568 (2007)

40. Rossi, R., Cavina, M., Setti, L.: Characterization of electron transfer mechanism in mediated microbial fuel cell by entrapped electron mediator in saccharomyces cerevisiae. Chem. Eng. Trans. **49**, 559–564 (2016)
41. Adebule, A.P., Aderiye, B.I., Adebayo, A.A.: Improving bioelectricity generation of microbial fuel cell (MFC) with mediators using kitchen waste as substrate. Ann. Appl. Microbiol. Biotechnol. J. **2**(1), 1–5 (2018)
42. Yifeng, Z., Liping, H., Jingwen, C., Xianliang, Q., Xiyun, C.: Electricity generation in microbial fuel cells: using humic acids as a mediator. J. Biotechnol. **136**, 474–475 (2008)
43. Xu, B., Ge, Z., He, Z.: Sediment microbial fuel cells for wastewater treatment: challenges and opportunities. Environ. Sci. Water Res. Technol. **1**(3), 279–284 (2015)
44. Strik, D., Timmers, R., Helder, M., Steinbusch, K., Hamelers, H., Buisman, C.: Microbial solar cells: applying photosynthetic and electrochemically active organisms. Trends Biotechnol. **29**(1), 41–49 (2011)

Chapter 2
Mathematical Modelling

A mathematical model is needed to represent the complex behavior of MFC into a set of simple mathematical formulations so as to optimally characterize the impact of operational and design constraints on the output [1–3]. Such mathematical modeling is done in two ways: engineering and statistical approaches as shown in Fig. 2.1. Ensemble models can also be formulated by integrating both engineering and statistical models.

The different operational parameters can be, bacterial growth rate, reaction rate, pH value, temperature, substrate concentration of the influent etc. while the design parameters are the surface area, sizes and materials (of electrodes), size of biofilm, electron donor variants, external resistance, membrane variants etc. MFCs are categorized into different types on the basis of chamber model, variants of microorganism cultures, substrate supply modes, and ion and electron transfer [4]. The various formulations of MFCs are shown in Fig. 2.2 for the ease of readers.

2.1 Engineering Based Modeling of MFCs

Engineering-based models are formulated based on the experimental study about the role of appropriate influencing parameters on output performance. They are represented by ordinary differential equations (ODEs) and partial differential equations (PDEs). An ODE model is mostly used to study the effect of substrate, biomass, by-product of MFCs operation, mediator concentration and charge balances. PDE based models are used to study the ion transportation through the membrane, biofilm growth at the electrode and substrate diffusion in the biofilm.

In [5], Zhang and Halme developed the simplified mathematical model of a one-dimensional two-chamber MFC with one anode, three parallel cathodes and three membranes separating the two chambers. The sequence of process involved in this type of model is that, first physical modeling of substrate concentration and redox reaction of the mediator using Monod kinetics is done, and then electrochemical

© Springer Nature Switzerland AG 2020
R. Patel et al., *Adaptive and Intelligent Control of Microbial Fuel Cells*, Intelligent Systems Reference Library 161,
https://doi.org/10.1007/978-3-030-18068-3_2

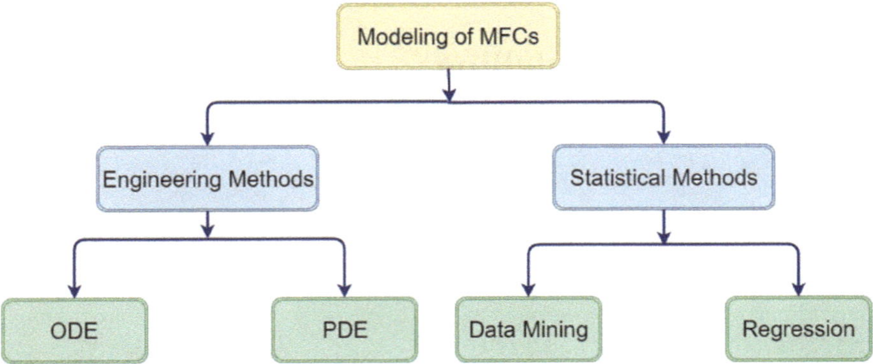

Fig. 2.1 Mathematical modeling approaches

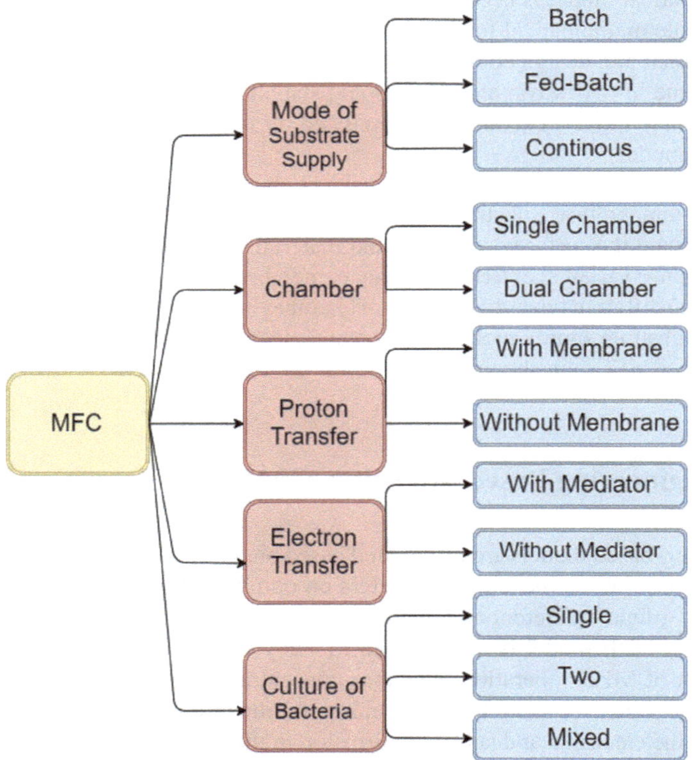

Fig. 2.2 Types of microbial fuel cells

modeling of current and electrode potentials using Faraday's law and Nernst's equation respectively is carried out to finally represent the MFC model in the form of ordinary differential equations. Experimental and simulation validation of output current of the model with different load conditions are given here.

In [6], Picioreanu et al. presented a modeling approach of batch mode two chamber MFC with G. Sulfurreducens as substrate and microorganism respectively. Mass balances of the substrate and biomass concentration are modeled using ODE approach and mass balance of biofilm is modeled using PDE approach. Butler-Volmer equation is used for the electrochemical modeling to calculate current density. They have also studied the impact of internal and external resistance on the output of MFCs. In [7], Zeng et al. have developed a mathematical model of continuous mode single bacterial species and dual chamber MFC. Substrate mass balance, biomass, and by-product of operation have been modeled using Monod kinetics and electrochemical modeling is done using the Butler-Volmer equation. They carried out two different experiments for the development of the model considering two distinct substrates such as acetate and glucose glutamic acid with a focus to model anode and cathode chamber reactions, but the biofilm growth cannot be studied here.

In [8], Pinto et al. presented a modeling of continuous mode, Single chamber, two population MFC with acetate as a substrate and two bacterial populations such as anodophilic and methanogenic bacteria. Physical modeling of mass balances of substrate, anodophilic, methanogenic concentration and intracellular mediator with Monod kinetics has been presented. Electrochemical modeling has been done by using ohm's law and Nernst's equation for the current and electrode potential calculation respectively. They developed four different MFCs test setups: two of those were used for estimating uncertain parameters, and the other two were used for model validation. In [9], Ravi Shankar et al. presented modeling and simulation of continuous mode, double chamber MFC with glucose glutamic acid as a substrate. This work mainly focused on the temperature variations with the current density in the cell chambers and heat transfer through the membrane.

In [10], Oliveira et al. proposed model of one-dimensional, two chamber MFC. The study mainly focused on the mass and heat transport at steady state condition using Fick models. Monod kinetics combined with Tafel equation is used for the anode modeling and only Tafel equation is used for the cathode modelling. In [11], Recio-Garrido et al. presented bioelectrochemical and electrical modeling of SC MFC with two bacterial population in two different approaches such as model-based simulation with known parameters (off-line approach) and model-based simulation with on-line estimation of physical and electrical parameters.

In [12], Abul et al. described a control-oriented model of SPSC MFC wherein acetate is used as a substrate and G. Sulfurreducens is the bacteria. A state representation of mass balance in anode chamber using Monod kinetics has been presented further. In [13], a mathematical modeling of batch mode, double chamber MFC with the pure Shewanella bacteria was developed using three different kinetics such as Monod, Blackman, and Tessier and further provided conclusion from their study that Monod kinetic is suited for the mass balances in each chamber in MFC model. Relevant equations pertaining to mathematical models of MFCs are in Table 2.1.

Table 2.1 Equations representing different types of MFCs models [14]

Nomenclature	Formulae	Application in models
Monod	$\mu = \mu_{max} \frac{C_s}{K_s + C_s}$	Bactrial growth and substrate oxidation
Tafel	$E = E_{eq} + \frac{RT}{(1-\alpha)nF} ln(\frac{i}{i_0})$	Electrode kinetics
Nernst	$E = E_0 - \frac{RT}{(1-\alpha)nF} ln(Q)$	Electrochemical behavior
Butler-Volmer	$i = i_0\left[exp(\alpha_a \frac{nF}{RT}(E - E_{eq})) - exp(-\alpha_c \frac{nF}{RT}(E - E_{eq})) \right]$	Current density

In these formulae, μ and μ_{max} are specific microorganism growth and the corresponding maximum growth respectively, C_s and K_s are the concentration of substrate and half-saturation constant respectively, E, E_{eq} and E^0 are the different potentials associated with electrode, equilibrium and standard electrode respectively, i and i_0 are the electrode current density A/m^2 and exchange current density A/m^2 respectively, F, T and R are Faraday's constant, absolute temperature and universal gas constant respectively, α_a and α_c are the dimensionless charge transfer coefficients (anode and cathode), n is the number of electrons involved in the electrode reaction, and Q is the reaction quotient.

For the reaction $xA + yB \rightarrow zC + wD$, Q is defined as

$$Q = \frac{[C]^z[D]^w}{[A]^x[B]^y}.$$

A detailed overview of different MFC models are provided in Table 2.2 depicting their features and characterization.

2.2 Mathematical Modelling of MFCs

Although substantial information presently exists from conventional fuel cells on mass transfer, electrical phenomena and reactions, the process of transfer of electrons from cells to electrode and on to the load, and the microbiological process of MFCs are not yet fully deciphered.

Development of MFC mathematical models based on deeper understanding of the involved processes, that allows one to reproduce the mathematics involved, can critically impact the continuous development of MFC technology and commercialization. The models presented herein are some basic steps towards this objective. The goal is to model the kinetics of current generation from an MFC inoculated with microbes. This section includes mathematical models of different MFCs based on chamber of operation and bacterial culture. These models are classified into mech-

Table 2.2 Modeling categorization of MFCs

Model approach	Compartment modeled	Mediator	No of bacterias	Reaction equations	Time and space resolution	References
ODE	Anode	Yes (External)	Single	Monod, Tafel, Nernst	1-D, Dynamic	[5]
		—	Single	Nernst, Monod	1-D, Dynamic	[12]
		Yes (Intra-cellular)	Two	Butler-Volmer, Double Monod, Nernst	1-D, Steady St., Dynamic	[8]
					1-D, Dynamic	[11]
	Anode, Cathode	No		Butler-Volmer, Monod	1-D,	[7]
		No		Tafel, Monod	Steady St., Dynamic	[10]
		Yes (External)	Single	Blackman, Monod, Tessier, Nernst		[13]
		No	Two	—	1-D, Steady St.	[9]
PDE	Anode	No	Single	Monod, Nernst	1-D, Steady St., Dynamic	[18]
					Dynamic 2-D, Steady St.,	[19, 20]
	Anode, Cathode			Butler-Volmer Monod, Nernst	1-D, Steady St.,	[21]
PDE&ODE	Cathode, Anode	Yes (External)	Single	Monod, Nernst	1-D, 2-D, 3-D, Steady St.	[22]
	Anode		Multiple	Butler-Volmer, Double Monod, Nernst	3-D, Steady St.	[23, 24]

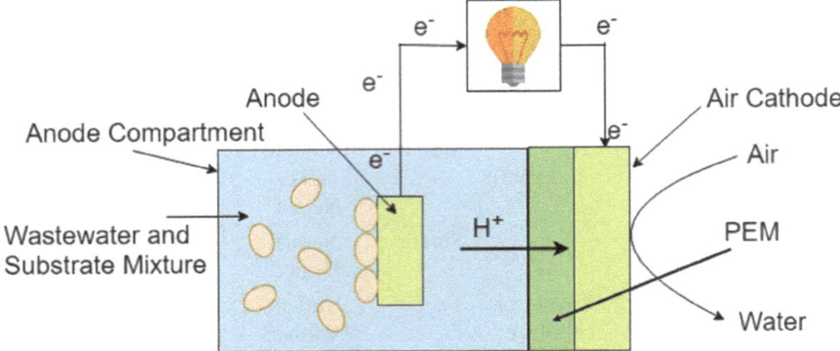

Fig. 2.3 SC MFC with PEM

anism based models which are mainly focused on the reaction process including substrate utilization, biofilm formation, bacterial or biomass growth etc..

2.2.1 *Single-Population Single Chamber MFC*

A typical MFC consists of electrodes (anode and cathode), food for microorganisms (substrate), cation ion membrane and microorganisms. The working mechanism and arrangement of the various sections in an MFC are shown in Fig. 2.3.

Microorganisms available in anode compartment produce positive ions (protons) and negative ions (electrons) by consuming substrates like nitrilotriacetic acid, acetate, ribitol, cysteine, glucose, lactate, glucuronic acid etc. [15]. Bacterias have a tendency to transfer generated electrons on anode surface in two ways: direct or indirect electron transfer. In the direct method, electrons are transferred through nanowires or mediators whereas external mediator is added in indirect electron transfer [16]. In the direct mode, electrons get transferred to cathode from the anode section via an external load [17], and the protons move to cathode. Electricity generation is achieved across the load and fresh water is obtained at cathode by combination of electron and proton. Anode based single species MFC with membrane developed by Ali Abul et al. is shown in Fig. 2.3 where G. sulfurreducens is used as the bacterium and substrate is the acetate. Assumptions include: ideal mixture of acetate, no further introduction of microbial biomass, minimal biofilm acetate gradient, and steady operational temperature [12].

The chemical reaction in both the chambers are

$$CH_3COO^- + 4H_2O \rightarrow 2HCO_3^- + 9H^+ + 8e^-,$$
$$O_2 + 4H^+ + 4e^- \rightarrow 2H_2O.$$

2.2.1.1   Microbial Kinetics

Active biomass and utilization of substrate are represented by Monod kinetics. The addition of the rates in microorganism synthesis and decay equals the net biomass growth rate given by

$$\mu = \left(\frac{1}{X}\frac{dX}{dt}\right)_{syn} + \left(\frac{1}{X}\frac{dX}{dt}\right)_{dec} = \mu_{syn} + \mu_{dec} = \mu_{max}\frac{C_s}{k_s + C_s} - K_d,$$

where C_S is the concentration of substrate, μ_{syn} and μ_{max} represent microbial growth rate and maximal growth rate respectively, X refers to the concentration of biomass, $K_d > 0$ is the decay coefficient, K_s is the half saturation constant.

Microorganisms have a tendency to break the substrate, oxidize it and use as food for living. The growth of microorganisms cell obtained from substrate dynamics is

$$q = q_{max}\frac{C_s}{K_s + C_s}X, \tag{2.1}$$

where q and q_{max} are the substrate utilization rate and its maximum rate respectively. The substrate utilization and microorganism biomass growth are linked by bacterial growth yield is given by

$$\mu_{max} = Yq_{max}.$$

2.2.1.2   Physical Model

The MFC system is fed by a feed flow of substrate with a rate of Q_a at anode compartment to govern the operations. The substrate and microorganism biomass dynamics are given by

$$\frac{dC_s}{dt} = -qX + D(C_{so} - C_s), \tag{2.2}$$

$$\frac{dX}{dt} = Yq_{max}\frac{C_s}{K_s + C_s}X - K_dX - DX, \tag{2.3}$$

where the system input is the dilution rate D, and Y refers to the bacterial yield [12]. All typical values of single species anode based microbial fuel cell is shown in Table 2.3.

Table 2.3 Nominal values: single species anode based MFC [12]

Symbol	Description	Nominal value	Unit
Y	Growth yield	0.11	Dimensionless
X_0	Initial biomass concentration	1.5	mgL^{-1}
q_{max}	Maximal substrate utilization rate	3	d^{-1}
μ_{max}	Maximal bacterial rate of growth	0.4	d^{-1}
K_d	Endogenous decay coefficient	0.084	d^{-1}
K_s	Half-saturation constant	32.4	mgL^{-1}
C_{so}	Influent substrate concentration	60	mgL^{-1}

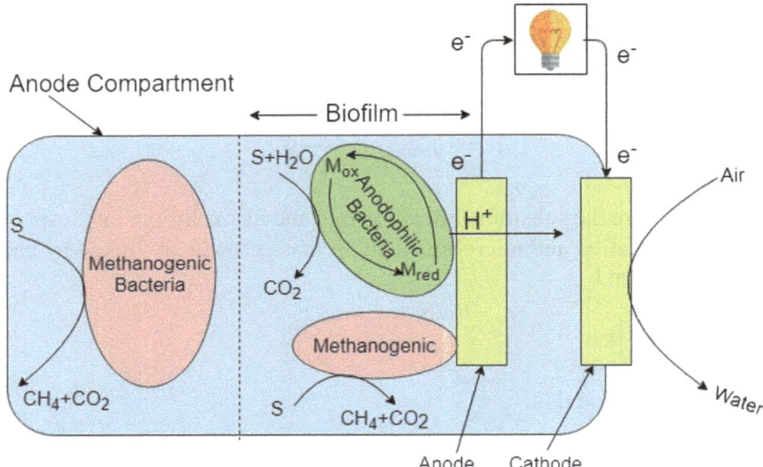

Fig. 2.4 SC two-population MFC without PEM

2.2.2 *Two-Population Single Chamber Microbial Fuel Cell*

Wastewater has diverse culture of microorganisms such as fermentative, methanogenic, and anodophilic [25]. The output of MFCs depends on the microorganism culture. Pinto et al. developed a dual species (anodophilic and methanogenic bacterias) anode based MFC, but bio-electrochemical reaction at cathode is not studied [8]. Two species anode based MFC sans membrane is shown in Fig. 2.4.

The performance is dependent upon the balance of the diverse microbial communities. Uniform bacterial species distribution is assumed in the anode compartment, and operational temperature and pH values are maintained constant. The reactions in anode chamber are

$$S + M_{ox} \rightarrow M_{red} + CO_2, \tag{2.4}$$

$$M_{red} \rightarrow M_{ox} + e^- + H^+, \tag{2.5}$$

$$S \rightarrow CH_4 + CO_2, \tag{2.6}$$

where S is the substrate concentration, M_{red} and M_{ox} represent the concentration of intracellular mediators in reduced and oxidized forms respectively.

The anode compartment mass balance equations are

$$\frac{dC_s}{dt} = -q_a X_a - q_m X_m + D(C_{so} - C_s), \tag{2.7}$$

$$\frac{dX_a}{dt} = \mu_a X_a - K_{d,a} X_a - \alpha_a D X_a, \tag{2.8}$$

$$\frac{dX_m}{dt} = \mu_m X_m - K_{d,m} X_m - \alpha_m D X_m, \tag{2.9}$$

$$\frac{dM_{ox}}{dt} = -Y q_a + \gamma \frac{I_{MFC}}{zF} \frac{1}{V_a X_a}, \tag{2.10}$$

where C_s and C_{so} are concentration of substrate (acetate) and influent substrate respectively, 'a' stands for anodophilic bacterias, 'm' refers to methanogenic bacterias and 'M' stands for mediator, X_a and X_m are the concentrations of microorganisms. Similarly, μ_a and μ_m are the growth rates of the microorganisms and also q_a and q_m are the specific consumption rates of both the bacterias, K_d, D, and Y are the decay rate of microorganisms, dilution rate, and mediator yield respectively, F, z, and γ are Faraday's constant, number of electrons transfered, and molar mass of mediator respectively, and α_a, α_m are dimensionless retention parameters.

Concentration of substrate and oxidized mediator formation are limiting factors for anodophilic growth whereas methanogenic bacterias are limited only by substrate. In (2.11)–(2.15), μ_a, μ_m, q_a and q_m are defined using Monod kinetics as

$$\mu_a = \mu_{max,a} \frac{C_s}{K_{s,a} + C_s} \frac{M_{ox}}{K_M + M_{ox}}, \tag{2.11}$$

$$\mu_m = \mu_{max,m} \frac{C_s}{K_{s,m} + C_s}, \tag{2.12}$$

$$q_a = q_{max,a} \frac{C_s}{K_{s,a} + C_s} \frac{M_{ox}}{K_M + M_{ox}}, \tag{2.13}$$

$$q_m = q_{max,m} \frac{C_s}{K_{s,m} + C_s}, \tag{2.14}$$

$$\alpha = \frac{1 + tanh[K_x(X_a + X_m - X_{max})]}{2}, \tag{2.15}$$

where μ_{max} and q_{max} are the bacterial maximum growth rate and the consumption rate of substrate, $K_{s,a}$, $K_{s,m}$ and K_M are the half-saturation constants, M_{ox} and M_{red} are the oxidized and reduced mediator fraction per anodophilic and methanogenic

Table 2.4 Nominal values: multi-population SC MFC [8]

Symbol	Description	Typical value	Unit
Y	Methane yield	22.75	mg-M mg-S^{-1}
$K_{d,a}$	Decay rate of anodophilic microbes	0.039	d^{-1}
$K_{d,m}$	Decay rate of methanogenic microbes	0.002	d^{-1}
K_M	Mediator half-rate constant	0.01	mg-ML^{-1}
K_x	Steepness factor	0.04	Lmg-x^{-1}
$K_{s,a}$	Half-rate constant of anodophilics	20	mg-S L^{-1}
$K_{s,m}$	Half-rate constant of methanogens	80	mg-S L^{-1}
z	No. of electron transferred	2	mole^{-1}mol$^{-1}_{mediator}$
γ	Mediator molar mass	663400	mg-M mol$^{-1}_{mediator}$

microorganisms respectively, K_x is the steepness factor, and X_{max} is the concentration of maximum achievable bacterial biomass.

The rate of dilution (D) is the ratio of rate of input flow of the substrate to the anode chamber volume. The growth rate of microorganisms, (μ) is controlled by the dilution rate [26]:

$$D = Q \, V_a^{-1}, \quad D > \max(\mu_{max,a}, \mu_{max,m}). \tag{2.16}$$

The parameter α ensures that the total microbial population do not exceed X_{max}. Biomass grows without limit when the dilution rate, $D < \min(\mu_{max,a}, \mu_{max,m})$. For stable equilibrium, $X_a + X_m = X_{max}$ is needed which is possible only when $0 < \alpha < 1$ and is ensured by K_x [25]. Typical values of two species anode compartment based MFC is provided in Table 2.4.

2.2.3 Two Chamber Single-Population Microbial Fuel Cell

Zeng et al. provide bio-electrochemical modeling of single species mediator less anode-cathode based MFC. The two chambers are segregated by the proton exchange membrane (PEM) to transfer protons to the Cathode. Oxidation of substrates takes place in the anode compartment and generates positive ions (protons). Saturated water is used as supply for the cathode compartment and the dissolved oxygen in that combine with transported protons (Fig. 2.5).

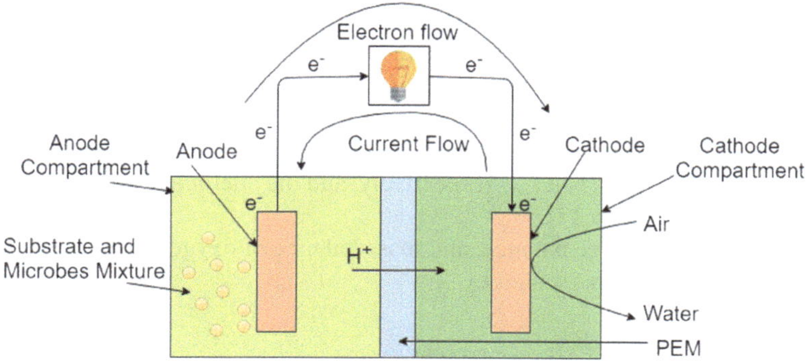

Fig. 2.5 Two chamber single-population MFC

A dynamic model of MFC is obtained through the combination of charge and mass balance, and bacterial kinetics. An anaerobic condition is made in anode compartment throughout the operation of MFC. Substrate is oxidized at the anode chamber and generates electrons and protons, which are transferred to the cathode chamber via an external circuit (load) and PEM respectively, as shown in (2.17) and (2.18). In cathode chamber, electron from anode combines with the saturated water. The protons don't involve in the cathode compartment chemical reaction [7]. The chemical reaction in both the compartments are represented as

$$(CH_2O)_2 + 2H_2O \rightarrow 2CO_2 + 8H^+ + 8e^-, \tag{2.17}$$

$$O_2 + 4e^- + 2H_2O \rightarrow 4OH^-. \tag{2.18}$$

At anode compartment, the charge balance, and mass balances of substrate, carbon dioxide, positive ions and active biomass of microorganisms are given by

$$V_a \frac{dC_{AC}}{dt} = Q_a(C_{AC}^{in} - C_{AC}) - A_m r_1, \tag{2.19}$$

$$V_a \frac{dC_{CO_2}}{dt} = Q_a(C_{CO_2}^{in} - C_{CO_2}) + 2A_m r_1, \tag{2.20}$$

$$V_a \frac{dC_H}{dt} = Q_a(C_H^{in} - C_H) + 8A_m r_1, \tag{2.21}$$

$$V_a \frac{dX}{dt} = Q_a \frac{(X^{in} - X)}{f_x} + A_m Y r_1 - V_a K_d X, \tag{2.22}$$

$$C_{cap,a} \frac{d\eta_a}{dt} = 3600 i_{cell} - 8F r_1. \tag{2.23}$$

In the anode chamber, the chemical reaction rate, r_1 is defined as

$$r_1 = K_1^0 exp\left(\frac{\alpha F}{RT}\eta_a\right) \frac{C_{AC}}{K_{AC} + C_{AC}} X. \tag{2.24}$$

The anode is indicated by the subscript 'a' and influent is identified in the equations by the superscript 'in', the forward rate constant related to anode reaction at standard condition is given by K_1^0, the surface area of membrane and wash-out fractional parameter respectively are represented by A_m and f_x, the universal gas constant, operational temperature, capacitance at anode, and current density of MFC are denoted as R, T, C_{cap} and i_{cell} respectively, and the coefficient of charge transfer at anode is indicated by α [7].

The equations of charge balance, and mass balance of oxygen, negative hydroxyl ions and positive ions represented by

$$V_c \frac{dC_{O2}}{dt} = Q_c(C_{O_2}^{in} - C_{O_2}) + A_m r_2, \tag{2.25}$$

$$V_c \frac{dC_{OH}}{dt} = Q_c(C_{OH}^{in} - C_{OH}) - 4A_m r_2, \tag{2.26}$$

$$V_c \frac{dC_M}{dt} = Q_c(C_M^{in} - C_M) + A_m N_M, \tag{2.27}$$

$$C_{cap,c} \frac{d\eta_c}{dt} = -3600 i_{cell} - 4F r_2. \tag{2.28}$$

The cathode compartment chemical reaction rate r_2 is defined as

$$r_2 = -K_2^0 exp\left(\frac{(\beta - 1)F}{RT}\eta_c\right)\frac{C_{O_2}}{K_{O_2} + C_{O_2}}. \tag{2.29}$$

The subscript, 'c' stands for the cathode, K_2^0 and β are forward rate constant and cathode charge transfer coefficient respectively, N_M is the M^+ ion flux moved from cathode chamber to anode via a membrane, expressed as

$$N_M = \frac{3600 i_{cell}}{F}. \tag{2.30}$$

In MFCs, microorganism growth depends on operational as well as environmental temperature, and best growth range is experimentally found to be 30°–40°C [27]. The influent concentration values except C_{AC}^{in} and $C_{O_2}^{in}$ are zero. The cathode charge transfer coefficient (β) is a significant parameter [7].

The nominal values of dual chamber anode-cathode based MFC model are provided in Table 2.5 and the performance is limited by cathodic reaction.

2.3 Control-Oriented Parametrized Models

The overall performance relies on certain attributes like substrate and biomass concentrations, growth rates, pH value, and operating and environmental temperature. However, the output voltage is affected by the aforementioned factors. The operation

Table 2.5 Nominal values of dual chamber MFC [7]

Symbol	Description	Typical value	Unit
A_m	Area of membrane Bacterial	5×10^{-5}	m^2
f_x	Reciprocal of wash-out fraction	10	–
Y	Bacterial yield	0.05	–
K_d	Decay constant for acetate utilizers	8.33×10^{-4}	h^{-1}
K_{AC}	Half velocity rate constant for acetate	0.592	$mol\ m^{-3}$
K_1^0	Forward rate constant of anode reaction at standard condition[a]	0.207	$mol\ m^{-2}\ h^{-1}$
C_{AC}^{in}	Influent substrate concentration in anode chamber	1.56	$mol\ m^{-3}$
K_2^0	Forward rate constant of cathode reaction at standard condition[a]	0.004	$mol\ m^{-3}$
$C_{O_2}^{in}$	Influent CO_2 concentration in anode chamber	0.3125	$mol\ m^{-3}$

[a]Standard Condition:- Absolute temperature $(T) = 303$ k, universal gas constant $(R) = 8.31447\ Jmol^{-1}k^{-1}$ and Pressure $(P) = 1$ bar

must be performed at controlled conditions so as to obtain a stable performance with optimal efficiency. Development of advanced methods for control and optimization strategy is required for a successful scale-up of the MFC technologies, and a favorable performance for various electric loads.

Bio-electrochemical models (discussed earlier) supplement the advanced control schemes. Control-oriented models of three different MFCs with uncertain parameters is provided where advanced control strategy can be easily applied. A backstepping control methodology with adaptation of uncertain parameters is presented from first principles for MFCs with one chamber and one type of species.

2.3.1   *Single-Population Single Chamber MFC*

Substrate concentration, C_s and biomass concentration, X are considered as x_1 and x_2. Dilution rate, D is considered as a input u. Maximum growth rate of microorganism μ_{max} represented as θ_1^{-1}, depends on the substrate and organisms type. The control-oriented parametrized mathematical model is

$$\dot{x}_1 = -\theta_1^{-1}.Y^{-1}\frac{x_1}{K_s + x_1}x_2 + u(C_{so} - x_1), \qquad (2.31)$$

$$\dot{x}_2 = \left(\theta_1^{-1}\frac{x_1}{K_s + x_1} - K_d - u\right)x_2. \qquad (2.32)$$

Abul et al. describes relevant system states, but since two of those states are minimal state space form of $\dot{x}_1$, one may neglect those two states for controller design. Values of q_{max} and μ_{max} are $3\ d^{-1}$ and $0.4\ d^{-1}$ respectively [12]. Maximal growth rate is based on the organism type and the limiting nutrients [26]. Pinto et al. suggest with 95% confidence the following parametric bound: $0.3699 < \theta_1^{-1} < 0.4301$ [8].

2.3.2 Two-Population Single Chamber MFC

The dynamics of MFC is given in (2.7)–(2.10) by several states, parameter (μ_{max}) and constants. Let us consider x_1, x_2, and x_3 be the concentrations of acetate (substrate), anodophilic and methanogenic microorganisms respectively, x_4 refers to the oxidized mediator fraction per anodophilic bacteria (M_{ox}). Dilution rate, D is considered as a manipulated input variable, u. The relationship between maximum acetate utilization rate and bacterial biomass growth rate is defined by $\mu_{max} = Y.q_{max}$ and it is also applicable to both the microorganisms. Maximum bacterial growth of anodophilic and methanogenic microorganisms ($\mu_{max,a}$, and $\mu_{max,m}$) are the uncertain parameters θ_1^* and θ_2^* respectively, and k_1 and k_2 are the inverse of such bacterial yields. The control-oriented parametrized mathematical model is

$$\dot{x}_1 = -k_1\theta_1^*\frac{x_1}{K_{s,a} + x_1}\frac{x_4}{K_M + x_4}x_2 - k_2\theta_2^*\frac{x_1}{K_{s,m} + x_1}x_3 + u(C_{so} - x_1),$$

$$\dot{x}_2 = x_2\left(\theta_1^*\frac{x_1}{K_{s,a} + x_1}\frac{x_4}{K_M + x_4} - K_{d,a} - u\alpha_a\right),$$

$$\dot{x}_3 = x_3\left(\theta_2^*\frac{x_1}{K_{s,m} + x_1} - K_{d,m} - u\alpha_m\right),$$

$$\dot{x}_4 = -\theta_1^*\frac{x_1}{K_{s,a} + x_1}\frac{x_4}{K_M + x_4} + \gamma\frac{I_{MFC}}{zF}\frac{1}{V_c x_2}. \qquad (2.33)$$

The effective control operation of microbial fuel cells need the parameters to be bounded within known intervals. Such regions for $\mu_{max,a}$ and $\mu_{max,m}$ are 3.01% and 3.4% respectively, and the values of $\mu_{max,a}$ and $\mu_{max,m}$ are $1.97\ d^{-1}$ and $0.1\ d^{-1}$ respectively. Pinto et al. have been carried out several experiments to find out the identifiable intervals with a level of confidence of 95% [8]. The parameters are bounded within these ranges: $1.939 < \theta_1^* < 2.000$, $0.06 < \theta_2^* < 0.134$.

2.3.3 Single-Population Two Chamber MFC

The performance of two chamber MFC largely relies on the chemical reactions in the chambers and the inputs. In anode chamber, the substrate (acetate) concentration C_{AC} is considered as a state x_1 and the other states x_2, x_3, x_4 and x_5 are defined as

$$x_2 = C_{AC} + \frac{1}{2}C_{CO_2}, \quad x_3 = C_{AC} + \frac{1}{8}C_H,$$
$$x_4 = YC_{AC} + X, \qquad x_5 = \zeta_1 C_{AC} - \eta_a,$$

where $\zeta_1 = 8FV_a/A_m$ is a constant. The dynamical equations representing control-oriented parametrized mathematical model of anode chamber is given by

$$\dot{x}_1 = -\frac{A_m}{V_a}K_1^0 exp\left(\delta_1^*(\zeta_1 x_1 - x_5)\right)\frac{x_1}{K_{AC} + x_1}(x_4 - Yx_1) + u_1(C_{AC}^{in} - x_1),$$

$$\dot{x}_2 = u_1\left(C_{AC}^{in} + \frac{1}{2}C_{CO_2}^{in} - x_2\right),$$

$$\dot{x}_3 = u_1\left(C_{AC}^{in} + \frac{1}{8}C_H^{in} - x_3\right),$$

$$\dot{x}_4 = Yu_1(C_{AC}^{in} - x_1) + \frac{u_1}{f_x}(X^{in} - x_4 + Yx_1) - K_d(x_4 - Yx_1),$$

$$\dot{x}_5 = \zeta_1 u_1(C_{AC}^{in} - x_1) - \frac{3600 i_{cell}}{C_{cap,a}},$$

where

$$\delta_1^* = \frac{\alpha F}{RT},$$

for known constants $\alpha = 0.051$, $F = 96485.4$ Coulombs mol^{-1}, $R = 8.3144$ J mol^{-1} K^{-1} and a range of temperature given by $303K < T < 313K$, which provides a range $1.890 < \delta_1^* < 1.9532$ [7]. The input u_1 is defined as the manipulated input variable of the anode chamber.

In cathode compartment, oxygen concentration, C_{O2} is considered as a state x_6 and the other states x_7, x_8 and x_9 are given as

$$x_7 = C_{O_2} + \frac{1}{4}C_{OH}, \quad x_8 = C_M, \quad x_9 = \zeta_2 C_{O_2} + \eta_c,$$

where $\beta_2 = 4FV_c/A_m$ is a constant. By using (2.25)–(2.28), The dynamics of control-oriented parametrized mathematical model of cathode chamber is represented by $\dot{x}_i$, $i = 6, 7..., 9$ as

$$\dot{x}_6 = u_2(C_{O_2}^{in} - x_6) - \frac{A_m K_2^0}{V_c} exp\left(\delta_2^*(x_9 - \zeta_2 x_6)\right)\frac{x_6}{K_{O_2} + x_6},$$

$$\dot{x}_7 = u_2\left(C_{O_2}^{in} - \frac{1}{4}C_{OH}^{in} - x_7\right),$$

$$\dot{x}_8 = u_2(C_M^{in} - x_8) + \frac{A_m N_M}{V_c}$$

$$\dot{x}_9 = \zeta_2 u_2(C_{O_2}^{in} - x_1) - \frac{3600 i_{cell}}{C_{cap,c}},$$

where u_2 is the input of the cathode chamber, $\delta_2^* = \frac{(\beta-1)F}{RT}$, and β is dependent on the input feed flow rate and decreases with decrease in the input flow rate. The proposed control-oriented models may be applied both in the on-line mode and in the off-line mode control strategy by choosing the voltage and substrate concentration as the output. Parameters of MFCs were estimated in on-line mode whereas the off-line mode required good knowledge and behavior of parameters to get better performance. Since $Q_a \in [2.25 \times 10^{-5} - 1.2 \times 10^{-5}]$ m^3 h^{-1}, therefore $0.663 < \beta < 1.367$, and $-12.90 \le \delta_2^* < 11.37$ [7].

This chapter discussed the chronological development of the mathematical models of MFCs using different approaches bringing out the analysis in terms of the number of microorganisms, chamber modelled, modelling perspective etc. Out of the several models discussed in this chapter control-oriented models are the most appropriate one for the development of an attractive renewable technology to provide efficient electrical energy and simultaneously also treats wastewater. Model based control and optimization techniques developed by identifying several parameters are found to vary within certain known bounds or rather ranges. The parametric variation in the models allows one to analyze the performance and behavior of the multifarious systems under uncertain conditions and make it a suitable topic of investigation for the development of advanced control strategies.

Further, as the main focus of this book is towards control-oriented models of MFCs so a detailed review is presented on the topic and subsequently following recommendations are made to conclude this chapter on modelling:

- It is necessary to understand the applications and operational requirements for choosing the type of MFCs such as SC or DC, and suitable mathematical models.
- It is advisable to identify the practical bounds of the uncertain parameters with higher percentage of confidence level through experiments.
- Valid modification in the mathematical model may be required to develop effective control and optimization techniques for enhancing overall performance.
- The application of control techniques should ensure that the operations of MFCs are environment friendly, cost effective and efficient.

References

1. Jadhav, G., Ghangrekar, M.: Performance of microbial fuel cell subjected to variation in pH, temperature, external load and substrate concentration. Bioresour. Technol. **100**, 717–723 (2009)
2. Kim, J., Cheng, S.: Membranes in microbial fuel cells. Environ. Sci. Technol. **41**, 1004–1009 (2007)
3. Logan, B., Hamelers, B., Rozendal, R., SchrÃuder, U., Keller, J., Freguia, S.: Microbial fuel cells: methodology and technology. Environ. Sci. Technol. **40**, 5181–5192 (2006)
4. Luo, S., Sun, H., Ping, Q., Jin, R., He, Z.: A review of modeling bioelectrochemical systems: engineering and statistical aspects. Energies **9**(2), 1–27 (2016)
5. Zhang, X., Halme, A.: Modelling of a microbial fuel cell process. Biotechnol. Lett. **17**, 809–814 (1995)
6. Picioreanu, C., Katuri, K., van Loosdrecht, M., Head, I., Scott, K.: Modelling microbial fuel cells with suspended cells and added electron transfer mediator. J. Appl. Electrochem. **40**, 151–162 (2009)
7. Zeng, Y., Choo, Y., Kim, B., Wu, P.: Modelling and simulation of two-chamber microbial fuel cell. J. Power Sour. **195**, 79–89 (2010)
8. Pinto, R., Srinivasan, B., Manuel, M., Tartakovsky, B.: A two-population bioelectrochemical model of a microbial fuel cell. Bioresour. Technol. **101**, 5256–5265 (2010)
9. Shankar, R., Mondal, P., Chand, S.: Modelling and simulation of double chamber microbial fuel cell: cell voltage, power density and temperature variation with process parameters. Green **3**, 181–194 (2013)
10. Oliveira, V., Simões, M., Melo, L., Pinto, A.: A 1D mathematical model for a microbial fuel cell. Energy **61**, 463–471 (2013)
11. Recio-Garrido, D., Perrier, M., Tartakovsky, B.: Combined bioelectrochemical Selectrical model of a microbial fuel cell. Bioprocess Biosystems Eng. **39**, 267–276 (2015)
12. Abul, A., Zhang, J., Steidl, R., Reguera, G., Tan, X.: Microbial fuel cells: control-oriented modeling and experimental validation. In: 2016 American Control Conference (ACC) (2016)
13. Esfandyari, M., Fanaei, M., Gheshlaghi, R., Akhavan Mahdavi, M.: Mathematical modeling of two-chamber batch microbial fuel cell with pure culture of Shewanella. Chem. Eng. Res. Design **117**, 34–42 (2017)
14. Ortiz-Martínez, V.M., Salar-García, M.J., de los Ríos A, Hernández-Fernández F.J., Egea J.A., Lozano L.: Developments in microbial fuel cell modeling. Chem. Eng. J. **271**, 50–60 (2015)
15. Pant, D., Van Bogaert, G., Diels, L., Vanbroekhoven, K.: A review of the substrates used in microbial fuel cells (MFCs) for sustainable energy production. Bioresour. Technol. **101**, 1533–1543 (2010)
16. Lovley, D.: The microbe electric: conversion of organic matter to electricity. Curr. Opin. Biotechnol. **19**, 564–571 (2008)
17. Lovley, D.: Bug juice: harvesting electricity with microorganisms. Nature Rev. Microbiol. **4**, 497–508 (2006)
18. Kato Marcus, A., Torres, C., Rittmann, B.: Conduction-based modeling of the biofilm anode of a microbial fuel cell. Biotechnol. Bioeng. **98**, 1171–1182 (2007)
19. Jayasinghe, N., Franks, A., Nevin, K., Mahadevan, R.: Metabolic modeling of spatial heterogeneity of biofilms in microbial fuel cells reveals substrate limitations in electrical current generation. Biotechnol. J. **9**, 1350–1361 (2014)
20. Merkey, B., Chopp, D.: The performance of a microbial fuel cell depends strongly on anode geometry: a multidimensional modeling study. Bull. Math. Biol. **74**, 834–857 (2011)
21. Sirinutsomboon, B.: Modeling of a membraneless single-chamber microbial fuel cell with molasses as an energy source. Int. J. Energy Environ. Eng. **5**(93) (2014)
22. Deb, D., Tao, G., Burkholder, J., Smith, D.: An adaptive inverse control scheme for a synthetic jet actuator model. In: Proceedings of the 2005, American Control Conference (2015)
23. Picioreanu, C., Head, I., Katuri, K., van Loosdrecht, M., Scott, K.: A computational model for biofilm-based microbial fuel cells. Water Res. **41**, 2921–2940 (2007)

24. Picioreanu, C., van Loosdrecht, M., Katuri, K., Scott, K., Head, I.: Mathematical model for microbial fuel cells with anodic biofilms and anaerobic digestion. Water Sci. Technol. **57**(7), 965–971 (2008)
25. Pinto, R., Tartakovsky, B., Perrier, M., Srinivasan, B.: Optimizing treatment performance of microbial fuel cells by reactor staging. Indu. Eng. Chem. Res. **49**, 9222–9229 (2010)
26. Maier, R., Pepper, I.: Environmental Microbiology. Elsevier, 3rd ed. (2015)
27. Naureen, Z., Rashid Al Matani, Z., Nasser Al Jabri, M., Al Housni, S., Abdullah Gilani, S., Mabood, F.: Generation of electricity by electrogenic bacteria in a microbial fuel cell powered by waste water. Adv. Biosci. Biotechnol. **07**, 329–335 (2016)

Chapter 3
Model Analysis of Single Population Single Chamber MFC

This chapter deals with the linearization of SPSC MFC model. The linearized system is analyzed with two distinct manipulated input variable cases: (1) dilution rate as a input, and (2) influent substrate concentration as a manipulated input variable. First, the equilibrium points are investigated and the stability at such points using Jacobian matrix is analyzed. An approximate linear model of the SPSC MFC is provided.

3.1 Introduction

In practical scenario, most of the systems contains nonlinearity terms. However, linear systems are easier to understand compared with nonlinear systems. Linearization of system provides a linear approximation of a nonlinear system around its operating region or equilibrium points. It is required to develop a classical control techniques and also helps to analyze the system stability and its behavior through root locus, bode plot, and Nyquist plot. It can be provided the knowledge about the system behavior around equilibrium points which are either stable or unstable and how the system will approach that points like move away or closer to the equilibrium points. The drawback of this model is that it is valid for small range of an equilibrium point.

First, identify the nonlinear components from the dynamical equations and make the nonlinear differential equations. The linearization is the expansion of the nonlinear differential equations into a Taylor series about the operating point and neglect the higher higher order terms which must be small, the only linear terms is present. Consider a two state system, whose dynamics are given as

$$\dot{z}_1 = \frac{dz_1}{dt} = f_1(z_1, z_2, u) \tag{3.1}$$

$$\dot{z}_2 = \frac{dz_2}{dt} = f_2(z_1, z_2, u) \tag{3.2}$$

$$y = h(z_1, z_2, u) \tag{3.3}$$

where z_1 and z_2, u, and y refer to the system states, control input, and output respectively, f and h refer to the nonlinear functions of z_1 z_2, and u. The objective is to convert nonlinear system into linear state space model, given as

© Springer Nature Switzerland AG 2020

R. Patel et al., *Adaptive and Intelligent Control of Microbial Fuel Cells*, Intelligent Systems Reference Library 161,
https://doi.org/10.1007/978-3-030-18068-3_3

$$\dot{Z} = AZ + Bu, \tag{3.4}$$

$$Y = CZ + Du. \tag{3.5}$$

For the linearization about the steady state, equilibrium points are obtained by taking (3.1) and (3.2) equal to zero.

$$f_1(z_{1s}, z_{2s}, u_s) = f_2(z_{1s}, z_{2s}, u_s) = 0 \tag{3.6}$$

where z_{is}, $i = 1, 2$ and u_s is a steady state equilibrium point and input. Perform Taylor series expansion of the above functions at the steady state operating points.

$$f_1(z_1, z_2, u) = f_1(z_{1s}, z_{2s}, u_s) + (z_1 - z_{1s})\frac{\partial f_1}{\partial z_1}\bigg|_{z_{1s}, z_{2s}, u_s} + (z_2 - z_{2s})\frac{\partial f_1}{\partial z_2}\bigg|_{z_{1s}, z_{2s}, u_s}$$

$$+(u - u_s)\frac{\partial f_1}{\partial u}\bigg|_{z_{1s}, z_{2s}, u_s} + higher\ order\ terms, \tag{3.7}$$

$$f_2(z_1, z_2, u) = f_2(z_{1s}, z_{2s}, u_s) + (z_1 - z_{1s})\frac{\partial f_2}{\partial z_1}\bigg|_{z_{1s}, z_{2s}, u_s} + (z_2 - z_{2s})\frac{\partial f_2}{\partial z_2}\bigg|_{z_{1s}, z_{2s}, u_s}$$

$$+(u - u_s)\frac{\partial f_2}{\partial u}\bigg|_{z_{1s}, z_{2s}, u_s} + higher\ order\ terms, \tag{3.8}$$

$$h_1(z_1, z_2, u) = h_1(z_{1s}, z_{2s}, u_s) + (z_1 - z_{1s})\frac{\partial h_1}{\partial z_1}\bigg|_{z_{1s}, z_{2s}, u_s} + (z_2 - z_{2s})\frac{\partial h_1}{\partial z_2}\bigg|_{z_{1s}, z_{2s}, u_s}$$

$$+(u - u_s)\frac{\partial h_1}{\partial u}\bigg|_{z_{1s}, z_{2s}, u_s} + higher\ order\ terms. \tag{3.9}$$

Neglecting the higher order terms and substituting the (3.6) in above equations. Note that the higher order terms should be small in value. The variables deviation from the steady state is defined as $Z = z_i - z_{is}$, $U = u - u_s$, and $Y = y - y_s z$. At the steady state,

$$\frac{dz_{is}}{dt} = f(z_{is}) = 0, i = 1, 2. \tag{3.10}$$

Therefore the $\frac{dz_i}{dt}$ is defined as

$$\frac{dz_i}{dt} = \frac{d(z_i - z_{is})}{dt} \tag{3.11}$$

Substituting the (3.11) into (3.7), (3.8), and (3.9), the system state space representation is provided as

$$\begin{bmatrix} \frac{d(z_1-z_{1s})}{dt} \\ \frac{d(z_2-z_{2s})}{dt} \end{bmatrix} = \begin{bmatrix} \frac{\partial f_1}{\partial z_1}\big|_{z_{1s}, z_{2s}, u_s} & \frac{\partial f_1}{\partial z_2}\big|_{z_{1s}, z_{2s}, u_s} \\ \frac{\partial f_2}{\partial z_1}\big|_{z_{1s}, z_{2s}, u_s} & \frac{\partial f_2}{\partial z_2}\big|_{z_{1s}, z_{2s}, u_s} \end{bmatrix} \begin{bmatrix} z_1 - z_{1s} \\ z_2 - z_{2s} \end{bmatrix} + \begin{bmatrix} \frac{\partial f_1}{\partial u}\big|_{z_{1s}, z_{2s}, u_s} \\ \frac{\partial f_2}{\partial u}\big|_{z_{1s}, z_{2s}, u_s} \end{bmatrix} \begin{bmatrix} u - u_s \end{bmatrix},$$

$$y - y_s = \left[\left.\frac{\partial h}{\partial z_1}\right|_{z_{1s},z_{2s},u_s} \quad \left.\frac{\partial h}{\partial z_2}\right|_{z_{1s},z_{2s},u_s} \right] \begin{bmatrix} z_1 - z_{1s} \\ z_2 - z_{2s} \end{bmatrix} + \left[\left.\frac{\partial h}{\partial u}\right|_{z_{1s},z_{2s},u_s} \right] \left[u - u_s \right].$$

The above equations are compared with (3.4) and (3.5) to obtain following matrices:

$$A = \begin{bmatrix} \left.\frac{\partial f_1}{\partial z_1}\right|_{z_{1s},z_{2s},u_s} & \left.\frac{\partial f_1}{\partial z_2}\right|_{z_{1s},z_{2s},u_s} \\ \left.\frac{\partial f_2}{\partial z_1}\right|_{z_{1s},z_{2s},u_s} & \left.\frac{\partial f_2}{\partial z_2}\right|_{z_{1s},z_{2s},u_s} \end{bmatrix}, \quad B = \begin{bmatrix} \left.\frac{\partial f_1}{\partial u}\right|_{z_{1s},z_{2s},u_s} \\ \left.\frac{\partial f_2}{\partial u}\right|_{z_{1s},z_{2s},u_s} \end{bmatrix},$$

$$C = \left[\left.\frac{\partial h}{\partial z_1}\right|_{z_{1s},z_{2s},u_s} \quad \left.\frac{\partial h}{\partial z_2}\right|_{z_{1s},z_{2s},u_s} \right], \quad D = \left[\left.\frac{\partial h}{\partial u}\right|_{z_{1s},z_{2s},u_s} \right].$$

The matrix A is called the Jacobian matrix. The system stability is facilitated by the negative eigenvalues of matrix A. The standard linearization procedure for two states for nonlinear dynamics and the control technique is described in [1–3]. Various control methods such as PI, PD, PID, linear quadratic control etc. can be applied to linearized model.

Consider a standard well known model of quadruple tank system. The system has four interconnected tanks and two pumps with inputs U_1 and U_2. The schematic description of the system is described in Fig. 3.1.

The control objective is to ensure a desired level of tank-1 and tank-2 which are considered as the outputs Y_1 and Y_2. Modeling of this system is presented in [4]. The nonlinear model of quadruple tank system is given as

$$\frac{dH_1}{dt} = \frac{1}{A_1}\left[\gamma_1 K_1 V_1 + a_3\sqrt{2gH_3} - a_1\sqrt{2gH_1} \right] = f_1(H_i, U_i)$$

$$\frac{dH_2}{dt} = \frac{1}{A_2}\left[\gamma_2 K_2 V_2 + a_4\sqrt{2gH_4} - a_2\sqrt{2gH_2} \right] = f_2(H_i, U_i)$$

$$\frac{dH_3}{dt} = \frac{1}{A_3}\left[(1 - \gamma_2) K_2 V_2 - a_3\sqrt{2gH_3} \right] = f_3(H_i, U_i)$$

$$\frac{dH_4}{dt} = \frac{1}{A_4}\left[(1 - \gamma_1) K_1 V_1 - a_4\sqrt{2gH_4} \right] = f_4(H_i, U_i)$$

$$Y_1 = K_c H_1 = h_1 \ , \quad Y_2 = K_c H_2 = h_2(H_i, U_i)$$

where A_i and a_i, $i = 1, 2, 3, 4$ refer to the cross section area of respective tanks and its orifice respectively, H_i refers to the water level of tanks, V_i, $i = 1, 2$, and k_i, $i = 1, 2$ refer to the input voltage, and flow constants of of pump-1 and pump-2 respectively, g represents the gravitational constant.

The nonlinear model is linearized using Taylor series approximation followed by Jacobian matrix transformation as per (3.7)–(3.9). The Jacobian matrices are defined as

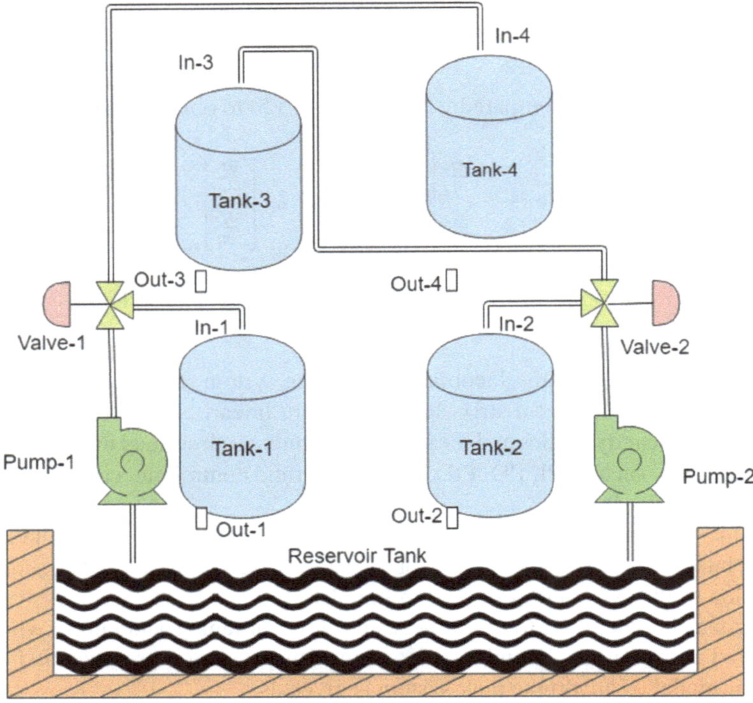

Fig. 3.1 Schematic diagram of the quadruple tank system

$$
A = \begin{bmatrix}
\frac{\partial f_1}{\partial H_1} & \frac{\partial f_1}{\partial H_2} & \frac{\partial f_1}{\partial H_3} & \frac{\partial f_1}{\partial H_4} \\[4pt]
\frac{\partial f_2}{\partial H_1} & \frac{\partial f_2}{\partial H_2} & \frac{\partial f_2}{\partial H_3} & \frac{\partial f_2}{\partial H_4} \\[4pt]
\frac{\partial f_3}{\partial H_1} & \frac{\partial f_3}{\partial H_2} & \frac{\partial f_3}{\partial H_3} & \frac{\partial f_3}{\partial H_4} \\[4pt]
\frac{\partial f_4}{\partial H_1} & \frac{\partial f_4}{\partial H_2} & \frac{\partial f_4}{\partial H_3} & \frac{\partial f_4}{\partial H_4}
\end{bmatrix}
=
\begin{bmatrix}
\frac{-1}{P_1} & 0 & \frac{A_3}{A_1 P_3} & 0 \\[4pt]
0 & \frac{-1}{P_2} & 0 & \frac{A_4}{A_2 P_4} \\[4pt]
0 & 0 & \frac{-1}{P_3} & 0 \\[4pt]
0 & 0 & 0 & \frac{-1}{P_4}
\end{bmatrix},
$$

$$
B = \begin{bmatrix}
\frac{\partial f_1}{\partial U_1} & \frac{\partial f_1}{\partial U_2} \\[4pt]
\frac{\partial f_2}{\partial U_1} & \frac{\partial f_2}{\partial U_2}
\end{bmatrix}
=
\begin{bmatrix}
\frac{\gamma_1 K_1}{A_1} & 0 \\[6pt]
0 & \frac{\gamma_1 K_1}{A_1} \\[6pt]
0 & \frac{(1-\gamma_2) K_2}{A_3} \\[6pt]
\frac{(1-\gamma_1) K_1}{A_4} & 0
\end{bmatrix},
$$

where P_i is defined as $\frac{a_i \sqrt{g}}{A_i \sqrt{2 H_i}}$, $i = 1, 2, 3, 4$. There are two outputs: level of the tank-1 and tank-2, and the system output matrix is defined as

$$
C = \begin{bmatrix}
\frac{\partial h_1}{\partial H_1} & \frac{\partial h_1}{\partial H_2} & \frac{\partial h_1}{\partial H_3} & \frac{\partial h_1}{\partial H_4} \\[4pt]
\frac{\partial h_2}{\partial H_1} & \frac{\partial h_2}{\partial H_2} & \frac{\partial h_2}{\partial H_3} & \frac{\partial h_2}{\partial H_4}
\end{bmatrix}
=
\begin{bmatrix}
K_c & 0 & 0 & 0 \\
0 & K_c & 0 & 0
\end{bmatrix},
$$

$$D = \begin{bmatrix} \frac{\partial h_1}{\partial U_1} & \frac{\partial h_1}{\partial U_2} \\ \frac{\partial h_2}{\partial U_1} & \frac{\partial h_2}{\partial U_2} \end{bmatrix} = \begin{bmatrix} 0 & 0 \\ 0 & 0 \end{bmatrix}.$$

The linearized state space representation of the quadruple tank system is provided in (3.4) and (3.5). The system state space model is the easiest or the most convenient option for the design of control algorithms and numerical calculations. One can convert this model into transfer functions based on requirements.

3.2 SPSC MFC Model with Dilution Rate as Input

SPSC MFC is described in the previous chapter. The performance of which is dependent on diverse factors like concentrations of biomass and substrate, pH value, operational temperature etc. The substrate and biomass concentrations are rely on the dilution rate and influent substrate concentration. Dilution rate, D ascertains how much media inflows the chamber on an hourly basis. For example, for a chamber volume of 200 mL^3 and a rate of dilution of 0.15, we get 30 mL media increment every hour.

The growth rate of bacterias is depended on the dilution rate and the different growth rates is possible to obtain by altering the rate of D. Note that, it is necessary to maintain the range of dilution rate and make sure that the range is within the limit provided by the maximum growth rate of bacterias [5]. The dilution rate can also affect on the characteristics of dynamic and steady state biofilm. The lower dilution rates increase the biomass production which leads to poor substrate diffusion via biofilm, whereas the higher rates leads to the rough structure and erosion of biofilm [6]. The dynamics of fermentation process is somewhat similar to the bacterial kinetics in MFC. The result of dilution rate on the enhancement of cells and productivity is extensively studied in [7–9].

The nonlinear mathematical model of SPSC MFC is given as

$$\dot{x}_1 = -\theta_1 k_1 \frac{x_1}{k_s + x_1} x_2 + u(S_0 - x_1) = f_1(x_1, x_2, u), \tag{3.12}$$

$$\dot{x}_2 = \left(\theta_1 \frac{x_1}{k_s + x_1} - k_d - u\right) x_2 = f_2(x_1, x_2, u), \tag{3.13}$$

where u is the dilution rate considered as a manipulated input. The other variables are described in previous chapter.

3.2.1 Equilibrium of the System

It is important to study the system dynamics of MFC for designing advanced control techniques. An important system attribute is the set of stable equilibrium points. Taking (3.12) and (3.13) equal to zero, there is a trivial equilibrium point

$$(x_1, x_2) \;=\; (S_0, 0), \tag{3.14}$$

where S_0 refers to the influent substrate concentration. The equilibrium point indicates that in absence of bacterial community, the concentration of the substrate is constant irrespective of input. The other nontrivial equilibrium dependent on input value is given by

$$(x_1 , x_2) \;=\; \left(\frac{k_s(k_d + u)}{\theta_1 - k_d - u} , \; \frac{u\,[s_0\theta_1 - (k_d + u)(k_s + S_0)]}{k_1(k_d + u)(\theta_1 - k_d - u)} \right). \tag{3.15}$$

From the equilibrium point, one can see that if $u = \theta_1 - k_d$, no finite equilibrium is possible.

3.2.2 Stability of Equilibrium Points

The stability of system is analyzed through the linearization of SPSC nonlinear model presented in (3.12) and (3.13). Perform the linearization using Taylor series approximation at the equilibrium points and the corresponding Jacobian matrix of the linearized system is given as,

$$A = \begin{bmatrix} \frac{\partial f_1}{\partial x_1} & \frac{\partial f_1}{\partial x_2} \\ \frac{\partial f_2}{\partial x_1} & \frac{\partial f_2}{\partial x_2} \end{bmatrix} = \begin{bmatrix} \frac{-\theta_1 k_1 x_2 k_s}{(k_s + x_1)^2} - u & \frac{-\theta_1 k_1 x_1}{k_s + x_1} \\ \frac{\theta_1 x_2 k_s}{(k_s + x_1)^2} & \frac{\theta_1 x_1}{k_s + x_1} - k_d - u \end{bmatrix},$$

$$B = \begin{bmatrix} \frac{\partial f_1}{\partial u} \\ \frac{\partial f_2}{\partial u} \end{bmatrix} = \begin{bmatrix} S_0 - x_1 \\ -x_2 \end{bmatrix}.$$

It can be seen that the system matrix A is depended on the manipulated input variable. The typical values parameters and constants of SPSC MFC model is given in Table 2.3. At the first trivial equilibrium point $(S_0, 0)$, the Jacobian matrix A is

$$A = \begin{bmatrix} -u & -2.36 \\ 0 & 0.175 - u \end{bmatrix}.$$

Note that, the above Jacobian matrix may stable for a certain range of input because the Eigenvalues are depended on the input variables. Eigenvalues of Jacobian matrix must be real and negative for guaranteed system stability. For the range of input $u < 0.175$, stable equilibrium point is established. The Eigenvalues of Jacobian

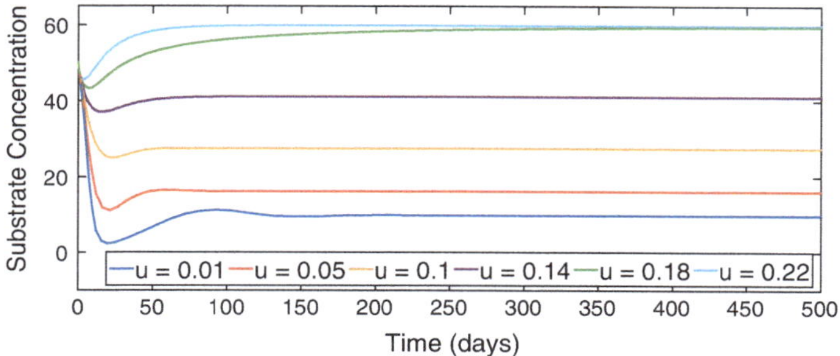

Fig. 3.2 Substrate concentration in nonlinear model with different inputs

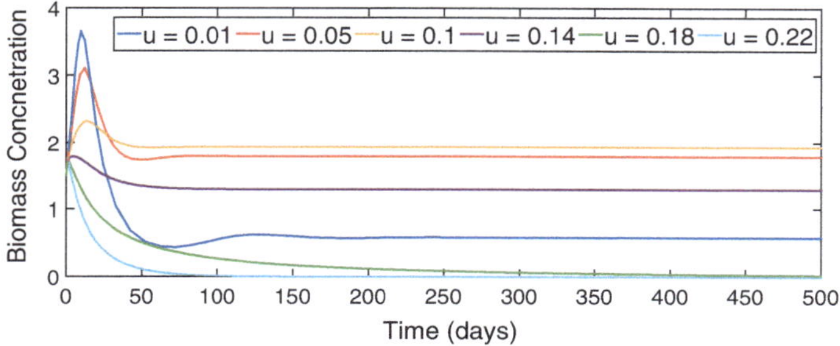

Fig. 3.3 Biomass concentration in nonlinear model with different inputs

matrix is negative for the input range, $u > 0.175$. The system is unstable for that particular range and needs suitable control techniques for desired stable performance.

Further system analysis has been made by applying step inputs with the range of $0.01 < u < 0.22$ to the nonlinear model of MFC and the effect of input on the substrate concentration and biomass concentration are described in the Fig. 3.2 and Fig. 3.3 respectively. It gives the information about changes and stabilization in concentrations with respect to the input. We obtain a range of input (D) $0.001 < D < 0.05$ for optimal growth of the bacteria in linearized system.

It can be observed that the biomass concentration is completely washout when the dilution rate is 0.18 and 0.22. At those input value, the substrate concentration is same as influent substrate concentration which shows no reactions happen in anode chamber. The effect of input on the substrate concentration and biomass concentration are in the Fig. 3.4 and Fig. 3.5 respectively.

In the linearized model analysis, the responses of the biomass and substrate concentrations with different dilution rate are almost same except the $u = 0.01$. Therefore, the performance of nonlinear and linearized system is almost similar at the dilution rate 0.01. The linearized system works fine around that particular input.

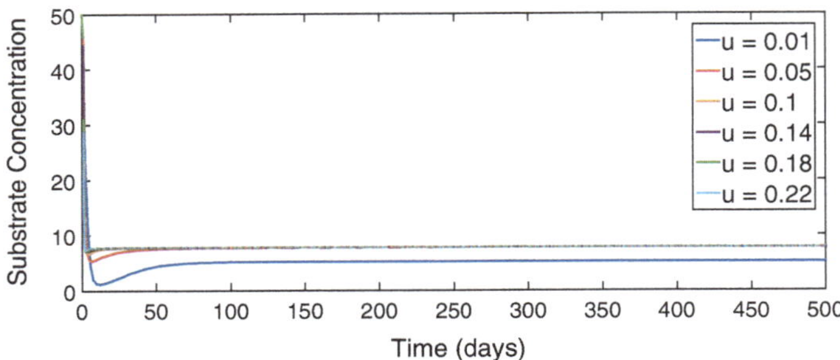

Fig. 3.4 Substrate concentration in linear model with different inputs

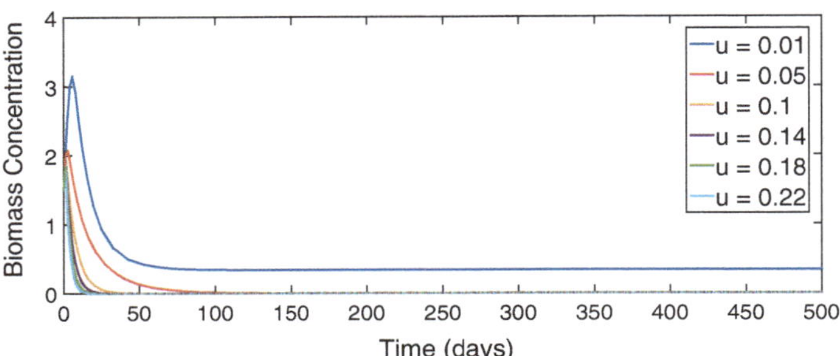

Fig. 3.5 Biomass concentration in linear model with different inputs

3.3   SPSC MFC Model with Influent Concentration as Input

The bacterial growth is highly depended on the substrate concentration and the specific substrate utilization rate is directly proportional to the specific bacterial growth rate [10]. The influent substrate concentration does affect the effluent quality and the concentration of microorganisms [11, 12]. The performance studies with the effect of influent concentration of biological wastewater treatment process is described [13].

The nonlinear mathematical model of SPSC MFC is given as

$$\dot{x}_1 \;=\; -\,\theta_1 k_1 \frac{x_1}{k_s + x_1} x_2 + D(u - x_1) \;=\; f_1(x_1, x_2, u), \tag{3.16}$$

$$\dot{x}_2 \;=\; \left(\theta_1 \frac{x_1}{k_s + x_1} - k_d - D\right) x_2 \;=\; f_2(x_1, x_2, u). \tag{3.17}$$

3.3.1 Equilibrium Points of the System

The nonlinear MFC model is provided in (3.16) and (3.17). SPSC MFC model can be studied in the fed-batch mode (input $u = 0$) and continuous mode (input u is variable). In fed-batch mode, the fresh media is added when the previous batch process is completed, while the fresh media is added continuously in continuous mode of operation [14]. Taking (3.16) and (3.17) equal to zero, there are two equilibrium points for continuous mode MFC shown as

$$(x_1,\ x_2) \;=\; (u, 0), \tag{3.18}$$

$$(x_1,\ x_2) \;=\; \left(p,\ \frac{d(u - p)(k_s + p)}{k_1 \theta_1 p} \right), \tag{3.19}$$

where p is defined as $\frac{k_s(k_d + d)}{\theta_1 - k_d - d}$. Both the equilibrium points are dependent on manipulated input variable (u). The equilibrium points of fed-batch mode MFC are

$$(x_1,\ x_2) \;=\; (0, 0), \tag{3.20}$$

$$(x_1,\ x_2) \;=\; \left(\frac{k_s(k_d + d)}{\theta_1 - k_d - d},\ \frac{-dk_s}{k_1(\theta_1 - k_d - d)} \right). \tag{3.21}$$

From the second equilibrium point, note that if the dilution rate, $d = \theta_1 - k_d$, there is no finite equilibrium achieved, and negative value of the biomass concentration is not possible. The biomass concentration is positive when $k_d + D > \theta_1$ and the equilibrium point may practically possible case for biomass concentration. However, it leads the negative substrate concentration which is not piratically possible Therefore, the second equilibrium point is not possible for the practical scenario.

3.3.2 Stability of the Equilibrium Points

Linearization of SPSC MFC has been done for fed-batch mode while considering input u as zero. Stability of linear system at equilibrium point is analyzed through Jacobian matrices which are given as

$$A = \begin{bmatrix} \frac{\partial f_1}{\partial x_1} & \frac{\partial f_1}{\partial x_2} \\ \frac{\partial f_2}{\partial x_1} & \frac{\partial f_2}{\partial x_2} \end{bmatrix} = \begin{bmatrix} \frac{-\theta_1 k_1 x_2 k_s}{(k_s + x_1)^2} - d & \frac{-\theta_1 k_1 x_1}{k_s + x_1} \\ \frac{\theta_1 x_2 k_s}{(k_s + x_1)^2} & \frac{\theta_1 x_1}{k_s + x_1} - k_d - d \end{bmatrix}, \quad B = \begin{bmatrix} \frac{\partial f_1}{\partial u} \\ \frac{\partial f_2}{\partial u} \end{bmatrix} = \begin{bmatrix} d \\ 0 \end{bmatrix}$$

The typical values parameters and constants of SPSC MFC model is given in Table 2.3. At the first equilibrium point $(0, 0)$, the Jacobian matrices are

$$A = \begin{bmatrix} -d & 0 \\ 0 & -k_d - d \end{bmatrix} = \begin{bmatrix} -0.09 & 0 \\ 0 & -0.174 \end{bmatrix}, \quad B = \begin{bmatrix} d \\ 0 \end{bmatrix} = \begin{bmatrix} 0.09 \\ 0 \end{bmatrix}$$

The eigenvalues of matrix A are -0.09 and -0.174 which indicate the linearized system is stable at equilibrium point $(0,0)$. At the second equilibrium point $\left(\frac{k_s(k_d + d)}{\theta_1 - k_d - d},\ \frac{-dk_s}{k_1(\theta_1 - k_d - d)} \right)$, the Jacobian matrices are

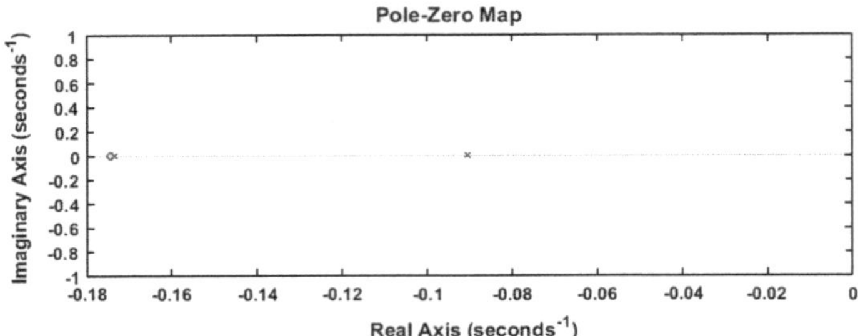

Fig. 3.6 Pole-zero plot of open-loop transfer function

$$A = \begin{bmatrix} -0.0785 & -1.58 \\ -0.0055 & -0.11 \end{bmatrix}, B = \begin{bmatrix} 0.09 \\ 0 \end{bmatrix}.$$

The eigenvalues of the matrix A are 0.0006 and -0.1886. The system is unstable at second equilibrium point due to one of the eigenvalue is positive. A suitable control technique such as pole placement control is necessary to make the system stable. The linearized system at equilibrium point $(0, 0)$ of SPSC MFC is

$$\dot{X} = AX + BU \tag{3.22}$$

where $X = [x_1 \; x_2]^T$, $U = [u \; 0]^T$, and A and B matrices are

$$A = \begin{bmatrix} -d & 0 \\ 0 & -k_d - d \end{bmatrix} = \begin{bmatrix} -0.09 & 0 \\ 0 & -0.174 \end{bmatrix}, B = \begin{bmatrix} d \\ 0 \end{bmatrix} = \begin{bmatrix} 0.09 \\ 0 \end{bmatrix}.$$

The equivalent open-loop transfer function of linearized system is obtained as

$$T(s) = \frac{0.09s + 0.0157}{s^2 + 0.2640s + 0.0157}. \tag{3.23}$$

The behavior and stability of the open-loop system is analyzed through bode-plot and pole-zero mapping. The pole-zero plot and bode plot of open-loop transfer function is shown in Fig. 3.6 and Fig. 3.7 respectively.

The pole-zero plot shows that all the poles are present in left side of s-plane which ensures system stability. Bode plot shows the phase margin of the system is $-180°$. The system gain is found to infinite, and so the system is always stable. It doesn't affect the system stability. It seems that the nature of the substrate and biomass concentration are approximately same in the linear and the nonlinear case. The linearization process is based on the approximation of nonlinear system, so it is difficult to get exact tracking of substrate and biomass concentrations in both the cases. The linearized fed-batch mode MFC model is used for the suitable control schemes for obtaining desired performance under controlled conditions. The open-loop simulation comparison of the nonlinear model and linear model at equilibrium point $(0,0)$ is shown in Figs. 3.8 and 3.9.

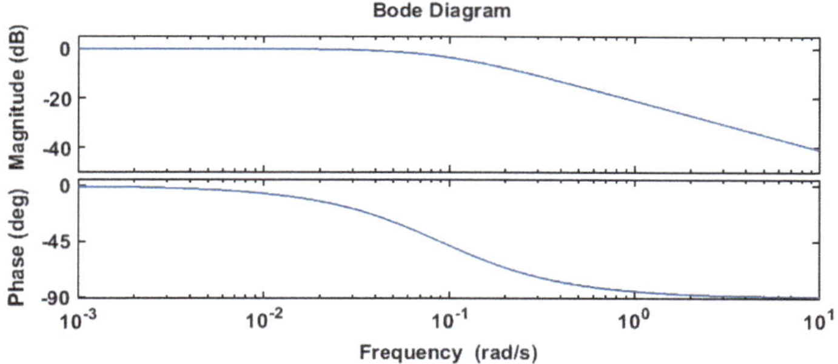

Fig. 3.7 Bode plot of open-loop transfer function

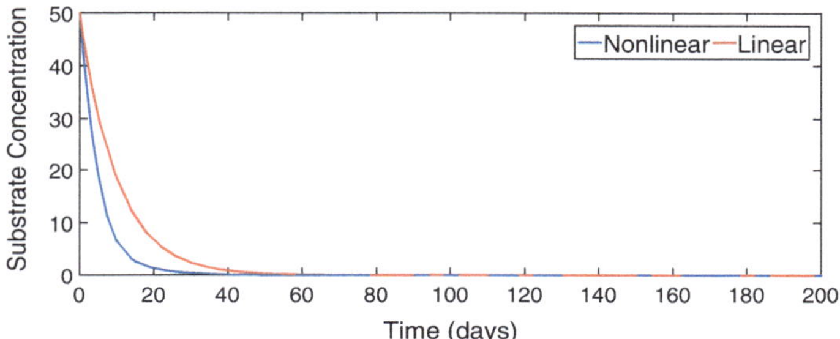

Fig. 3.8 Comparison of substrate concentration

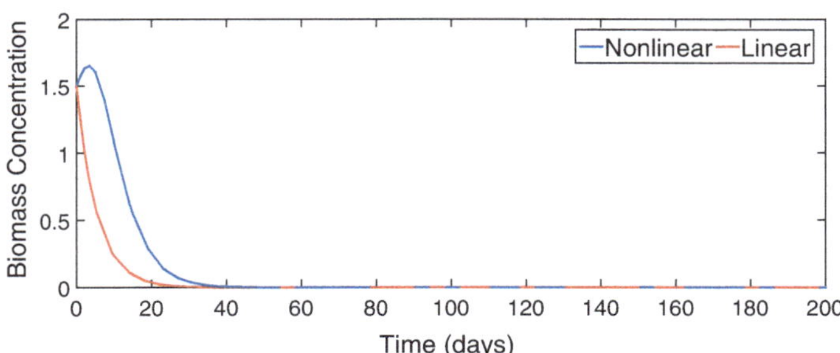

Fig. 3.9 Comparison of biomass concentration

This chapter discusses the linearization of SPSC MFC. The first part of the chapter deals with the fundamentals of linearization of nonlinear systems with suitable example. There are two general steps to find linearized model of any nonlinear system, equilibrium points and its stability. The second part of this chapter discusses the linearization of the SPSC MFC model with two different manipulated input variables. A detailed analysis of SPSC MFC with influent substrate concentration as a manipulated input variable is presented with stability analysis of equilibrium points and obtained open-loop transfer function. Further, this transfer function of MFC can use to develop various linear control techniques.

References

1. ElMaraghy, H., Lahdhiri, T., Ciuca, F.: Robust linear control of flexible joint robot systems. J. Intell. Robot. Syst. **34**(4), 335–356 (2002)
2. Guarnizo, M., Trujillo, R., Guacaneme, M.: Modeling and control of a two DOF helicopter using a robust control design based on DK iteration. In: IECON 2010-36th Annual Conference on IEEE Industrial Electronics Society (2010). https://doi.org/10.1109/iecon.2010.5675183
3. Jordan, A.: Linearization of non-linear state equation. Bull. Pol. Acad. Sci. **54**(1), 63–73 (2006)
4. Jayaprakash, J., Kumar, H.: State Variable analysis of four tank system. In: 2014 International Conference on Green Computing Communication and Electrical Engineering (ICGCCEE) (2014)
5. Ziv, N., Brandt, N., Gresham, D.: The use of chemostats in microbial systems biology. J. Vis. Exp. **80** (2013). https://doi.org/10.3791/50168
6. Rangappa, V., Kodialbail, V., Bharthaiyengar, S.: Effect of dilution rate on dynamic and steady-state biofilm characteristics during phenol biodegradation by immobilized Pseudomonas desmolyticum cells in a pulsed plate bioreactor. Front. Environ. Sci. Eng. **10**(4) (2016). https://doi.org/10.1007/s11783-016-0863-9
7. Szewczyk, K.: The effect of dilution rate variation on the performance of continuous fermentation. Bioprocess Eng. **6**, 17–19 (1991). https://doi.org/10.1007/bf00369273
8. Martens, D., de Gooijer, C., van der Velden-de, Groot C., Beuvery, E., Tramper, J.: Effect of dilution rate on growth, productivity, cell cycle and size, and shear sensitivity of a hybridoma cell in a continuous culture. Biotechnol. Bioeng. **41**(4), 429–439 (1993). https://doi.org/10.1002/bit.260410406
9. Nabergoj, D., Kuzmić, N., Drakslar, B., Podgornik, A.: Effect of dilution rate on productivity of continuous bacteriophage production in cellstat. Appl. Microbiol. Biotechnol. **102**(8), 3649–3661 (2018). https://doi.org/10.1007/s00253-018-8893-9
10. Kayombo, S., Mbwette, T., Katima, J., Jorgensen, S.: Effects of substrate concentrations on the growth of heterotrophic bacteria and algae in secondary facultative ponds. Water Res. **37**(12), 2937–2943 (2003). https://doi.org/10.1016/s0043-1354(03)00014-9
11. Grady, C., Harlow, L., Riesing, R.: Effects of growth rate and influent substrate concentration on effluent quality from chemostats containing bacteria in pure and mixed culture. Biotechnol. Bioeng. **14**(3), 391–410 (1972). https://doi.org/10.1002/bit.260140310
12. Sánchez, E., Borja, R., Travieso, L., Martín, A., Colmenarejo, M.: Effect of influent substrate concentration and hydraulic retention time on the performance of down-flow anaerobic fixed bed reactors treating piggery wastewater in a tropical climate. Process. Biochem. **40**(2), 817–829 (2005). https://doi.org/10.1016/j.procbio.2004.02.005
13. Grady, C., Williams, D.: Effects of influent substrate concentration on the kinetics of natural microbial populations in continuous culture. Water Res. **9**(2), 171–180 (1975). https://doi.org/10.1016/0043-1354(75)90006-8
14. Pannell, T., Goud, R., Schell, D., Borole, A.: Effect of fed-batch versus continuous mode of operation on microbial fuel cell performance treating biorefinery wastewater. Biochem. Eng. J. **116**, 85–94 (2016)

Chapter 4
Robust Control Design of SPSC Microbial Fuel Cell with Norm Bounded Uncertainty

In this chapter an attempt is made for the first time to formulate a robust controller in a linear matrix inequality (LMI) framework for a linearized model of Single Population Single Chamber Microbial Fuel Cell (SPSC MFC) to improve the system performance with unstructured time-varying uncertainties. The dilution rate is considered as an uncertain parameter.

4.1 Introduction

MFC dynamic models are nonlinear in nature due to complex growth rate of bacterias. MFC performance is widely dependent on certain parameters such as concentrations of substrate and biomass, operational temperature, bacterias growth rates, pH and dilution rate etc. Nonlinear control methods will give better performance compared to linear controllers. However, It may be costlier or sometimes difficult to implement in real time and also increases complexity of the system. The conventional linear controllers such as PI, PID or optimal control techniques are designed for the nominal system at an operating point. However, these control methods cannot give satisfactory performance for time-varying parameters. Therefore, robust control methods such as H_∞, LMI etc. are needed to get desired performance against time-varying parameters. Several robust control methods using H_∞ or LMI framework are designed for linear systems [1–5].

The advantage of using LMI is that it provides a lucid analytical framework for controller design [6] and the LMI conditions can be solved easily with existing LMI toolbox of MATLAB [7]. A linearization of nonlinear SPSC MFC model at equilibrium point is presented in Chap. 3. The open-loop simulation comparison of the nonlinear model and linear model at equilibrium point (0, 0) is shown in Figs. 3.8 and 3.9. It seems that the nature of substrate and biomass concentration are approximately same in linear and nonlinear case. The linearized model is given as

$$\dot{X} = AX + BU, \tag{4.1}$$

where $X = [x_1, \ x_2]^T$, $U = [u, \ 0]^T$, and A and B matrices are

© Springer Nature Switzerland AG 2020

R. Patel et al., *Adaptive and Intelligent Control of Microbial Fuel Cells*, Intelligent Systems Reference Library 161,
https://doi.org/10.1007/978-3-030-18068-3_4

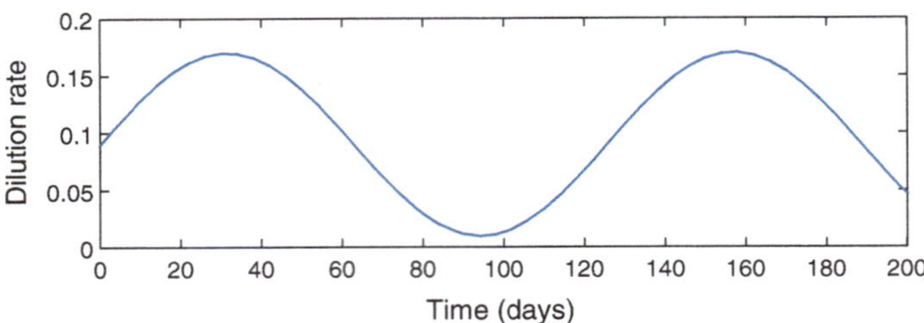

Fig. 4.1 Uncertain dilution rate with time

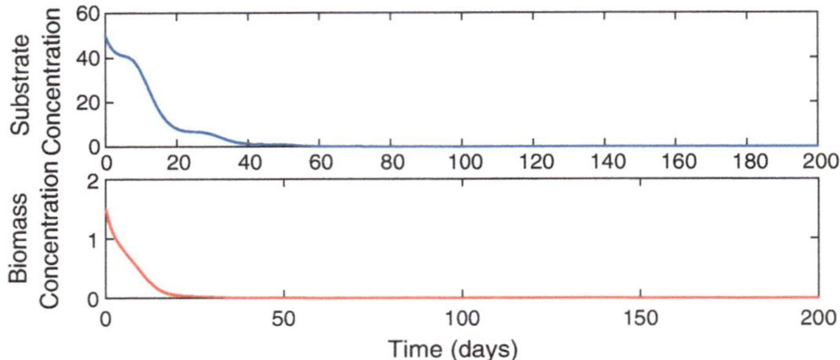

Fig. 4.2 Effect of dilution rate on substrate and biomass concentration

$$A = \begin{bmatrix} -d & 0 \\ 0 & -k_d - d \end{bmatrix} = \begin{bmatrix} -0.09 & 0 \\ 0 & -0.174 \end{bmatrix}, B = \begin{bmatrix} d \\ 0 \end{bmatrix} = \begin{bmatrix} 0.09 \\ 0 \end{bmatrix}.$$

The dilution rate, d in the anode chamber when controlled, allows one to control the bacterial growth [8]. In practice, the anode compartment volume is fixed, and so the dilution rate is only dependent on the feed flow rate. MFC is a bioelectrochemical process related to bacterial reactions with substrate. The reaction time is slow and also depends on several factors like environmental effect, pH values and concentration of substrate etc. The effect of dilution rate on microbial community is discussed [9].

If the dilution rate is uncertain due to some physical issue with the device, the MFC system performance gets affected. If the frequency of uncertainty in dilution rate is fast, it will not affect performance because the bacterial reaction process is slow but when the frequency of the uncertainty is slow and remains so for a long time, it affects the performance of MFC. Typical value of dilution rate (d) is 0.09 m^3d^{-1}. Consider dilution rate with sinusoidal uncertainty with time is defined such as $d \times \delta \sin(\omega t)$. The uncertain dilution rate at 0.05 Hz, as $\delta = 0.08$ and its effect on substrate and biomass concentration are shown in Fig. 4.1 and Fig. 4.2 respectively.

At very low frequency of uncertainty in dilution rate, the performance of MFC gets degraded and becomes very sluggish.

As the dilution rate is uncertain, the linear system has uncertainty in system matrix, A and input matrix, B. Robust controller with norm bounded uncertainty is needed to improved system performance against uncertainty in the dilution rate using LMI approach.

4.2 Brief Overview of LMI

The linear matrix inequalities (LMI) in control systems uses convex optimization theory that was initially developed from the linear programming using interior-point method. Later all the control problems were translated into set LMI constraints. The chapter will first introduce a standard definition of LMI for the ease of understanding of readers and will further proceed to highlight some interesting control problem that can be recast as an LMI problem. One of the most striking feature of LMI is that it can handle multi-objective optimization scenario with constraints. The standard definition of LMI is stated below mathematically and is taken from [10]:

$$\phi(y) = \phi_0 + \sum i y_i \phi_i > 0$$

where y is a vector and treated as a variable and $\phi_i = \phi_p^T$ is a square symmetric matrix. The above inequality means,

1. The matrix $\phi(y)$ is positive definite.
2. It is a set of n polynomial inequalities in the variable y.
3. Multiple LMIs can be expressed as,

$$\begin{bmatrix} \phi_1(y) & \dots & 0 \\ 0 & & \dots & 0 \\ 0 & \dots & 0 \end{bmatrix} > 0.$$

Symmetrical matrices Φ_i of the LMI set is given as

$$\phi = \left\{ y \in \mathbb{R}^n | \Phi(y) = \Phi_0 + \sum_{i=1}^n y_i \Phi_i > 0 \right\}.$$

If the diagonal minors $\phi_i(y)$ are positive then only the matrix $\Phi(y)$ is positive semi-definite and they are multivariate polynomials of independent y_i. Therefore the LMI set is given as

$$\Phi(x) = \left\{ y \in \mathbb{R}^n | \phi_i(y) > 0, i = 1, ..., n \right\}.$$

The definition of Semi-algebraic sets are sets which are defined by polynomial inequalities and equations. An example of the geometry of an LMI feasible set is taken from [11]:

$$\Phi(y) = \begin{bmatrix} 3 & 0 & 1 \\ 0 & 4 & 0 \\ 1 & 0 & 0 \end{bmatrix} + y_1 \begin{bmatrix} -1 & -1 & 0 \\ -1 & 0 & 0 \\ 0 & 0 & -1 \end{bmatrix} + y_2 \begin{bmatrix} 0 & -1 & 0 \\ -1 & -1 & 0 \\ 0 & 0 & 0 \end{bmatrix} > 0.$$

Equivalence of above function is given as

$$\Phi(y) = \Phi(y) = \begin{bmatrix} 3 - y_1 & -(y_1 + y_2) & 1 \\ -(y_1 + y_2) & 4 - y_2 & 0 \\ 1 & 0 & -y_1 \end{bmatrix} > 0.$$

The above function is feasible, if and only if, all the principal minors are non-negative system of polynomial inequalities, $\phi_i(y)$.

1. First order minors

$$\phi_1(y) = 3 - y_1 > 0$$
$$\phi_2(y) = 4 - y_2 > 0$$
$$\phi_3(y) = -y_1 > 0$$

2. Second order minors

$$\phi_4(y) = (3 - y_1)(4 - y_2) - (y_1 + y_2)^2 > 0$$
$$\phi_5(y) = -y_1(3 - y_1) - 1 > 0$$
$$\phi_6(y) = -y_1(4 - y_2) > 0$$

3. Third order minors

$$\phi_7(y) = -y_1((3 - y_1)(4 - y_2) - (y_1 + y_2)^2) - (4 - y_2) > 0.$$

In 1980's, there are two main advancement happened in LMI approach. The first one was an interior-point algorithm [12] which is a polynomial time linear programming technique and it is far better than the simplex technique in speed. The second advance was the convex optimization problems [13]. In late 1980's, researchers started to formulate control systems as a convex optimization problem with LMI approach and provided solutions using linear programming [14] and it was noticed that the convex optimization was considered as an emerging method for controller development [15].

LMI are used in System and Control and its applications for over a century. A practical LMI techniques are used in aircraft design which gives a suitable framework for aircraft controller development [16]. H_∞/H_2 controller design with LMI approach are developed for different applications such as satellite controller, space launch vehicle, compact disc player, automatic landing system of aircraft [17–20]. LMI approach was successfully implemented on fly by wire helicopter controller and also provide detailed robustness analysis of fighter aircraft control laws [21, 22]. Robotics world also used control techniques with LMI approach [23]. LMI techniques provide improved results over conventional control methods in different applications. In certain cases, there are numerical limitation with LMI solvers. However, many research work will be done to improve the system performance and robustness. For the first time the authors uses the LMI for formulating robust control problem of MFCs.

4.2.1 *Some Control Problems in LMI Framework*

Lyapunov stability is used for analysis of continuous linear time invariant system. The Lyapunov inequality is given as

$$A^T P + PA < 0, \ P > 0. \tag{4.2}$$

For the stable system $\dot{z} = Az$, the matrix P must be positive definite for the below linear equation.

$$A^T P + PA = -Q \tag{4.3}$$

Lyapunov noticed that the choosing any value of Q, $Q = Q^T > 0$ provides the first LMI solutions. The above concept is also formulated as a LMI optimization problem. Therefore, consider a system with the initial value $z(0) = z_0$

$$\dot{z}(t) = Az(t) \tag{4.4}$$

The criterion function is given as

$$J = \int_0^\infty z^T(t) Q z(t) dt \tag{4.5}$$

where matrix Q is positive definite and $Q = Q^T$. Take an assumption that the above system is asymptotically stable. Therefore, the solutions of the systems are bounded and also the $J < \infty$. Thus, a quadratic Lyapunov function is defined to ensure a boundedness on J. The quadratic Lyapunov function is given as

$$V(z(t)) = z^T(t) P z(t), \tag{4.6}$$

For some matrix $P = P^T > 0$, the condition is defined as

$$\frac{dV}{dt}(z(t)) = \frac{d}{dt}\left(z^T(t) P z(t)\right) \leq -z^T(t) Q z(t) dt \tag{4.7}$$

For all trajectories and time, t, the above condition is negative definite. Integrate the inequality form time $0 to T$.

$$z^T(T) P z(T) - z^T(0) P z(0) \leq -\int_0^T z^T(t) Q z(t) dt, \tag{4.8}$$

As long as $z^T(t) P z(t) \geq 0$, and it holds for $t \to \infty$, then

$$J = \int_0^\infty z^T(t) Q z(t) dt \leq z^T(0) Q z(0). \tag{4.9}$$

The boundedness of the criterion function is given as

$$J \leq z_0^T(t) Q z_0(t) \tag{4.10}$$

For any solution $P = P^T$ of (4.7), and the derivative is defined as

$$\frac{d}{dt}\left(z^T(t) P z(t)\right) = z^T(t)(A^T P + P A) z(t) \leq -z^T(t) Q z(t) \tag{4.11}$$

The above equation gives the equivalent condition as follows

$$A^T P + P A + Q < 0 \tag{4.12}$$

LMI feasibility problem is solved by obtaining a bound on J, which will be obtained by finding P. To obtain best bound on J, one can use optimization over the Lyapunov function P such as minimize $z_0^T P z_0$ subject to $P > 0$, $A^T P + P A + Q \leq 0$. $z_0^T P z_0$ is a linear function of the variable P and it is numerically tractable by using interior-point methods.

Schur Complements are used to convert nonlinear inequalities into LMI form. The basics of Schur Complement is given by following Lemma. The positive definite block matrix is given as

$$\begin{bmatrix} Q & S \\ S^T & R \end{bmatrix}$$

if and only if

$$Q > 0 \,\&\, R - S^T R^{-1} S > 0, \tag{4.13}$$

if and only if,

$$R > 0 \,\&\, Q - S R^{-1} S^T > 0, \tag{4.14}$$

For the state feedback control problem for the system described by

$$\dot{x}(t) = A x(t) + B u(t) \ , \ u(t) = K x(t) \tag{4.15}$$

where x and u refer to the system state and input signal respectively. The closed-loop system is given as

$$\dot{x}(t) = (A + B K) x(t) \tag{4.16}$$

So the Lyapunov expression for the stability of the closed-loop system is given as

$$P(A + B K) + P(A + B K)^T < 0 \tag{4.17}$$

Expanding the above equation and we obtain

$$P A^T + P A + P K^T B^T + B K P < 0 \tag{4.18}$$

The above equation is not LMI as the terms P and K matrices are unknown, thus once can use change of variable here. Let consider $KP = W$ and the above equation is written as

$$PA^T + PA + W^T B^T + BW < 0 \tag{4.19}$$

After solving this LMI, we will get the solution of P and W. Therefore, the gain matrix K can be calculated as,

$$K = WP^{-1} \tag{4.20}$$

4.2.2 LMI Solvers

There are many computationally efficient LMI solvers are available now for solving the sets of LMIs like, LMI toolbox of MATLAB, LMI-tool, YALMIP, CVX etc. Out of these available software LMI toolbox of MATLAB is used more often. The LMI formulation of the control problems is grouped into three classes:

1. Feasibility problem: Finding if there exists a solution of an LMI ($F(x) < 0$). The MATLAB function used to solve this type problem is **feasp** using LMI toolbox of MATLAB.
2. Optimization problem: Minimizing any linear objective $f(x)$ subject to an LMI constraint ($F(x) < 0$). The MATLAB function used to solve this type problem is **mincx** using LMI toolbox of MATLAB.
3. Generalized eigenvalue problem: Minimizing Eigenvalue λ subject to $G(x) - \lambda F(x) < 0$, $F(x) > 0$ and $H(x) < 0$. The MATLAB function used to solve this type problem is **gevp** using LMI toolbox of MATLAB.

4.3 Linear MFC Model with Uncertain Dilution Rate

Let us consider a linearized SPSC MFC model with time-varying norm bounded matrix uncertainty associated with the plant and input. The state-space form is

$$\dot{x}(t) = A(t)\,x(t) + B(t)\,u(t), \tag{4.21}$$

where $A(t)$ and $B(t)$ are time-varying and uncertain plant and input matrices respectively. These matrices are represented as

$$A(t) = A + \Delta A(t), \quad B(t) = B + \Delta B(t) \tag{4.22}$$

where A and B are nominal matrices and the time-varying norm bounded uncertainty is represented as $||\Delta A(t)|| = ||\Delta B(t)|| < d$ for a nominal value of dilution rate d, and $\Delta A(t)$, $\Delta B(t)$ is given as

$$\Delta A(t) = \begin{bmatrix} \delta \sin(\omega t) & 0 \\ 0 & \delta \sin(\omega t) \end{bmatrix}, \quad \Delta B(t) = \begin{bmatrix} \delta \sin(\omega t) \\ 0 \end{bmatrix}.$$

The above matrices are decomposed as

$$\Delta A(t) = D_a F_a(t) E_a, \quad \Delta B(t) = D_b F_b(t) E_b \tag{4.23}$$

where $F_a(t)$ and $F_b(t)$ are uncertain time-varying matrices lying in Lebesgue space, satisfying the condition $F_a^T(t) F_a(t) \leq I$ or $F_b^T(t) F_b(t) \leq I$, and D_a, D_b, E_a, E_b are the known constants that characterized how they affect the nominal system. The decomposition of the uncertain matrices are used to exploit the structure observed in the uncertainty and for normalization of components which are time-varying in nature [24].

4.4 Controller Design

The MFC plant with a feedback signal $u(t) = Kx(t)$ is given as

$$\dot{x}(t) = (A(t) + B(t)K)x(t). \tag{4.24}$$

The control objective is to effectively choose a gain K by using the Lemmas:

Lemma 4.1 ([25]) *Given matrices $Q = Q^T$, H, E and $R = R^T > 0$ of suitable dimensions, and $Q + HFE + E^T F^T H^T < 0$, $\forall\ F$ that satisfies $F^T F \leq R$, if and only if $\exists$ some $\lambda > 0$, such that $Q + \lambda H H^T + \lambda^{-1} E^T R E < 0$.*

Lemma 4.2 ([25] (Schur Complement)) *For any matrices Q, R and S*

$$\begin{bmatrix} Q & S \\ S^T & R \end{bmatrix} < 0$$

is equal to $R < 0$ and $Q - SR^{-1}S^T < 0$.

Theorem *If a matrix $Y = Y^T > 0$ and positive scalar $\epsilon_1 > 0$, $\epsilon_2 > 0$ exists and the LMI condition,*

$$\Lambda = \begin{bmatrix} YA^T + A^T Y + X^T B + BX + \epsilon_1 DD^T + \epsilon_2 DD^T & Y^T E_a^T & X^T E_b^T \\ * & -\epsilon_1 I & 0 \\ * & 0 & -\epsilon_2 I \end{bmatrix}$$

is satisfied then the system (4.24) has a stabilizing solution $K = YX^{-1}$ for all admissible uncertainties defined in (4.23). $\qquad\qquad\nabla$

Proof Lyapunov second method is adopted for designing the robust controller for the system in (4.24).

$$V(t) = x^T(t) P x(t). \tag{4.25}$$

The time derivative of (4.25) is

$$\dot{V}(t) = \dot{x}^T(t)Px(t) + x^T(t)P\dot{x}(t). \tag{4.26}$$

Substituting $\dot{x}(t)$ from (4.24) in (4.26), we get

$$\begin{aligned}
\dot{V}(t) &= [(A(t)+B(t)K)x(t)]^T Px(t) + x^T(t)P[(A(t)+B(t)K)x(t)] \\
&= x^T(t)A^T(t)Px(t) + x^T(t)K^T B^T(t)Px(t) + x^T(t)PA(t)x(t) \\
&\quad + x^T(t)PB(t)Kx(t). \tag{4.27}
\end{aligned}$$

Substituting $A(t)$ and $B(t)$, (4.27) can be written as

$$\begin{aligned}
\dot{V}(t) &= x^T(t)[A^T P + PA + K^T B^T P + PBK + E_a^T F^T D^T P + PDFE_a + PDFE_b K \\
&\quad + K^T E_b^T F^T D^T P]x(t). \tag{4.28}
\end{aligned}$$

One can re-write (4.28) according to Lemma 4.1 as

$$\begin{aligned}
\dot{V}(t) &\le x^T(t)[A^T P + PA + K^T B^T P + PBK + \epsilon_1 PD(PD^T) + \epsilon_1^{-1} E_a^T E_a \\
&\quad + \epsilon_2 PD(PD^T) + \epsilon_2^{-1} K^T E_b^T E_b K]x(t). \tag{4.29}
\end{aligned}$$

According to Lyapunov's theory $\dot{V}(t) < 0$ if the matrix Ω is negative definite, then

$$\dot{V}(t) \le x^T(t)\Omega x(t), \tag{4.30}$$

where $\Omega = [A^T P + PA + K^T B^T P + PBK + \epsilon_1 PD(PD^T) + \epsilon_1^{-1} E_a^T E_a + \epsilon_2 PD(PD^T) + \epsilon_2^{-1} K^T E_b^T E_b K]$. Through pre and post multiplication of Ω by P^{-1} and a linear change of variables by letting $Y = P^{-1}$ and $KY = X$ and an LMI criteria $\Lambda < 0$ for guaranteed asymptotic stability and admissible uncertainty.

The SPSC MFC model as in (4.21) with nominal parametric values is

$$A = \begin{bmatrix} -0.09 & 0 \\ 0 & -0.174 \end{bmatrix}, B = \begin{bmatrix} 0.09 \\ 0 \end{bmatrix}$$

The decomposition matrices pertaining to the uncertain set of parameters are

$$D_a = D_b = \begin{bmatrix} 0.282 & 0 \\ 0 & 0.282 \end{bmatrix}, E_a = \begin{bmatrix} 0.282 & 0 \\ 0 & 0.282 \end{bmatrix}, E_b = \begin{bmatrix} 0.282 \\ 0 \end{bmatrix}.$$

The closed loop simulation has been done for SPSC MFC model with uncertainty in dilution rate as discussed in Sect. 8.1. The uncertain model with robust controller using LMI approach has been simulated for two different frequencies of uncertainty

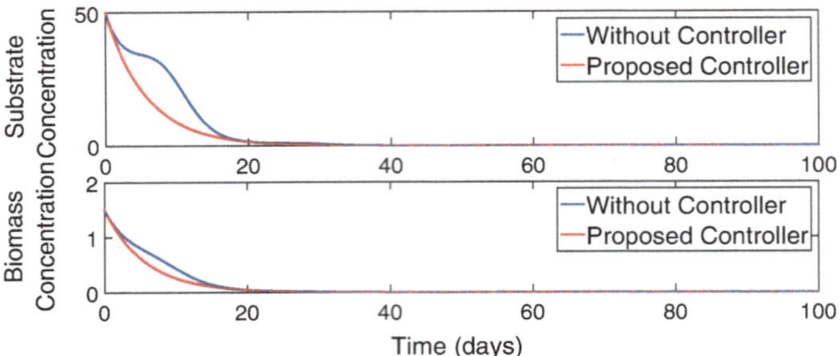

Fig. 4.3 Performance comparison at $f = 0.05$ Hz

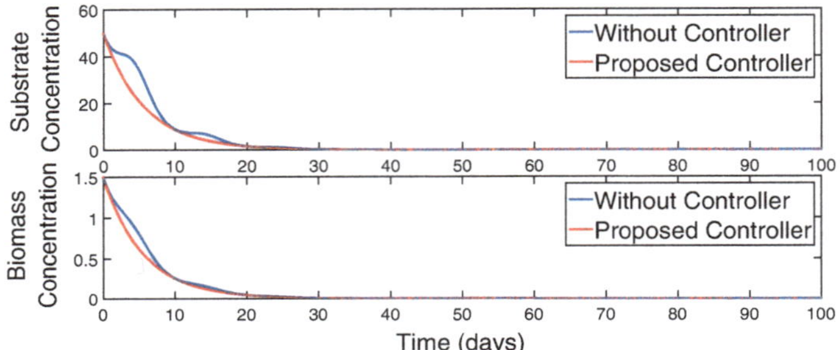

Fig. 4.4 Performance comparison at $f = 0.1$ Hz

with $f = 0.05$ Hz and $f = 0.1$ Hz. The performance of proposed controller is compared with uncertain system without controller with two different cases shown in Figs. 4.3 and 4.4.

The robust controller using LMI approach against norm bounded uncertainty in dilution rate provides good performance. The performance of system without controller is degraded with the change in frequency of uncertainty. Proposed controller provides robustness against the change in frequency of uncertainty in dilution rate.

This chapter dealt with the development of robust controller of SPSC MFC with norm bounded uncertainty in dilution rate. The first part of the chapter discussed the effect of uncertain dilution rate on performance and development of linear MFC model. Model based optimized controller is developed for appropriate behavior and performance analysis of MFC systems. The second part of the chapter dealt with a robust controller designed for SPSC MFC with norm bounded uncertainty in dilution rate using LMI framework which improves the controller performance with robustness against uncertainty in system and input matrices. The simulation work provides controller effectiveness and system performance under uncertainty in dilution rate.

References

1. Lai, C., Fang, C., Kau, S., Lee, C.: Robust H/sub 2/control of norm-bounded uncertain continuous-time system-an LMI approach. In: IEEE International Conference on Robotics and Automation, pp. 243–248 (2004)
2. Wang, G., Zhang, Q., Sreeram, V.: Robust H_∞ control of norm bounded uncertain systems via Markovian approach. Asian J. Control. **13**(6), 956–965 (2010)
3. Xie, L., de Souza, C.E.: Robust control for linear time-invariant systems with norm-bounded uncertainty in the input matrix. Syst. Control. Lett. **14**(5), 389–396 (1990)
4. Devarakonda, N., Yedavalli, R.K.: A new robust control design for linear systems with norm bounded time varying real parameter uncertainty. In: ASME 2010 Dynamic Systems and Control Conference, vol. 1, pp. 1–7 (2010)
5. Liu, Y., Lin, H., Liu, L., Li, Y.: A parameter-dependent approach to robust h control of norm bounded uncertain systems. Appl. Mech. Mater. 645–653 (2015)
6. Boyd, S., Ghaoui, L.E., Feron, E., Balakrishnan, V.: Linear Matrix Inequalities in System and Control Theory. Society for Industrial and Applied Mathematics (SIAM), Philadelphia (1994)
7. Gahinet, P., Nemirovski, A., Laub, A.J., Chilali, M.: LMI Control Toolbox: For Use with MATLAB. The MathWorks Inc, Natick, MA (1995)
8. Maier, R., Pepper, I.: Environmental Microbiology, 3rd edn. Elsevier (2015)
9. Winkler, M., Boets, P., Hahne, B., Goethals, P., Volcke, E.: Effect of the dilution rate on microbial competition: r-strategist can win over k-strategist at low substrate concentration. PLoS ONE (2017)
10. Stephen, B., Laurent, G., Eric, F., Venkataramanan, B.: Linear Matrix Inequalities in System and Control Theory. Society for Industrial and Applied Mathematics, Philadelphia (1994)
11. Parrillo, P.A., Lall, S.: Semidefinite programming relaxations and algebraic optimization in control. Eur. J. Control. **9**(2–3), 307–321 (2003)
12. Karmarkar, N.: A new polynomial-time algorithm for linear programming. Combinatorica **4**(4), 373–395 (1984)
13. Horisberger, H.P., Belanger, P.R.: Regulators for linear time-invariant plants with uncertain parameters. IEEE Trans. Autom. Control. **21**, 705–708 (1976)
14. Bernussou, J., Peres, P.L.D., Geromel, J.C.: A linear programming oriented procedure for quadratic stabilization of uncertain systems. Syst. Control. Lett. **13**(1), 65–72 (1989)
15. Boyd, S., Balakrishnan, V., Barratt, C., Khraishi, N., Li, X., Meyer, D.G., Norman S.A.: A new CAD method and associated architectures for linear controllers. IEEE Trans. Autom. Control. **33**(3), 268–283 (1988)
16. Niewoehner, R.J., Kaminer, I.I.: Integrated aircraft-controller design using linear matrix inequalities. J. Guid. Control. Dyn. **19**(2), 445–452 (1996)
17. Zasadzinski, M., Frapard, B.: Multiobjective controller designs: a space application benchmark. In: Workshop on Linear Matrix Inequalities in Control, Toulouse, France, LAAS-CNRS (2004)
18. Clement, B.: Robust control with LMI optimisation in space applications. In: Workshop on Linear Matrix Inequalities in Control, Toulouse, France, LAASCNRS (2004)
19. Scherer, C., Gahinet, P., Chilali, M.: Multiobjective output-feedback control via LMI optimization. IEEE Trans. Autom. Control **42**(7), 896–911 (1997)
20. Seto, D., Ferriera, E., Marz, T.: Case Study: Development of a Baseline Controller for Automatic Landing of an F-16 Aircraft Using Linear Matrix Inequalities (LMIs). Carnegie Mellon, Software Engineering Institute: 155: Research Report CMU/SEI-99-TR-020 (2000)
21. Dettori, M., Scherer, C.W.: MIMO control design for a compact disc player with multiple norm specifications. IEEE Trans. Control. Syst. Technol. **10**(5), 635–645 (2002)
22. Prempain, E., Postlethwaite, I.: Static H-infinity loop shaping control of a fly-by-wire helicopter. In: Workshop on Linear Matrix Inequalities in Control, also in Proceedings of the 43rd IEEE Conference on Decision and Control, Toulouse, France, LAAS-CNRS (2004)
23. Biannic, J.M.: IQC for robustness analysis of fighter aircraft control laws. In: Workshop on Linear Matrix Inequalities in Control, Toulouse, France, LAASCNRS (2004)

24. Dey, R., Ghosh, S., Ray, G.: A robust H_∞ load-frequency controller design using LMIs. In: 2009 IEEE International Conference on Control Applications (2009)
25. Dey, R., Ghosh, S., Ray, G., Rakshit, A., Balas, V.: Improved delay-range-dependent stability analysis of a time-delay system with norm bounded uncertainty. ISA Trans. **20**(58), 50–57 (2015)

Chapter 5
Introduction to Adaptive Control

Basic concepts of adaptive control and some standard adaptive control techniques are presented herein. These include direct and indirect adaptive control, adaptive backstepping control, and model reference adaptive control techniques. Each method is described from fundamentals to application in real-time systems.

5.1 Introduction

Plants with uncertainties requires highly efficient control schemes so as to achieve good performance. Adaptive control is a mechanism used when the system encounters uncertainties due to its ability to adapt or handle parameter uncertainties [1]. Adaptive control techniques have seen significant development [2]. Adaptive control techniques have proven effective for such systems. Meaning of the adaptive term in control prospective is to adapt or to change such that the system's behavior will conform to the changed condition. An adaptive controller consists of the parameter estimator and control law. The basic configuration of adaptive controllers with adaption scheme is described in Fig. 5.1.

The parameter estimator estimates the unknown or uncertain parameter. Adaptive control provides automatic adjustment of controllers for maintaining desired system performance when system parameters are not precisely known [3–7]. The performance of the system is measured and compared with the desired values. Based on error generated from comparison, the adjustable controller adapt the changed condition through different adaption mechanisms. Various adaptive control techniques are available such as direct and indirect adaptive control, adaptive backstepping control, MRAC, and adaptive pole placement control etc. Such control techniques are also developed and further analysis has taken place specifically to bio processes [8–11].

© Springer Nature Switzerland AG 2020

R. Patel et al., *Adaptive and Intelligent Control of Microbial Fuel Cells*, Intelligent Systems Reference Library 161,
https://doi.org/10.1007/978-3-030-18068-3_5

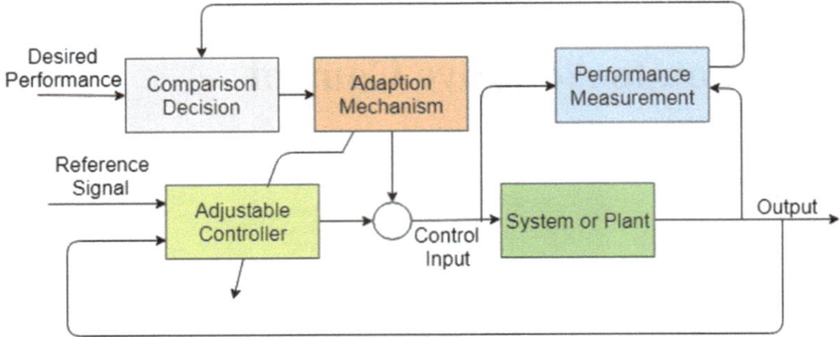

Fig. 5.1   Basic adaptive control configuration

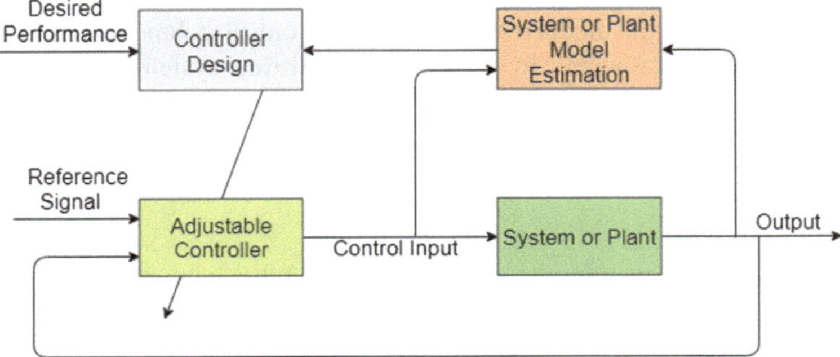

Fig. 5.2   Basic configuration of indirect adaptive control

5.2   Indirect Adaptive Control

In the indirect (or explicit) method, the adaptive controller parameters are indirectly
calculated based on the on-line estimated plant parameters [4]. The basic objective
of this control method is the estimation of the plant from possible input-output measured quantities. The adaptation mechanism has two different steps: on-line system
parameter estimation, and on-line calculation of the parameters of controller based
on model estimation. This scheme provides different control laws and estimation
methods. A basic block diagram is presented in Fig. 5.2.

Example of Indirect Adaptive Control:- Consider a scalar system

$$\dot{p} = ap + bu \tag{5.1}$$

where p and u refer to system state and input variable. However, a and b are unknown.
The control objective is to have the actual system follow the chosen reference system
given as

$$\dot{p}_r = a_r p_r + b_r r \tag{5.2}$$

where $p_r \in L_\infty$ and $r(t) \in L_\infty$ refer to reference system's state and input signal respectively. Note that, a_r must be negative. The reference model is selected in such a manner that desired closed-loop tracking is accomplished.

Since the system parameters are uncertain, it is difficult to compute controller gains k_p and k_r. Therefore, an adaptive controller is formulated as follows

$$u(t) = \hat{k}_p p + \hat{k}_r r. \tag{5.3}$$

Assuming that certain matching conditions given by

$$a + bk_p = a_r, \quad bk_r = b_r, \tag{5.4}$$

are satisfied, the estimates of a and b can be represented as $\hat{a}(t)$ and $\hat{b}(t)$ respectively. Therefore, the controller's gains and control input are represented as

$$\hat{k}_r = \frac{a_r - \hat{a}}{\hat{b}} \; , \quad \hat{k}_r = \frac{b_r}{\hat{b}}$$

$$u = \frac{a_r - \hat{a}}{\hat{b}} p + \frac{b_r}{\hat{b}} r. \tag{5.5}$$

Substituting (5.5) into (5.1) and get

$$\dot{p} = ap + b.\left(\frac{a_r - \hat{a}}{\hat{b}}\right) p \tag{5.6}$$

The state error and parameter errors are defined as $e = p - p_r$ and $\tilde{a} = a - \hat{a}, \tilde{b} = b - \hat{b}$ respectively. The derivative of state error is obtained as

$$\begin{aligned}
\dot{e} &= ap + bu - a_r p_r - b_r r \\
&= (a - \hat{a})p + a_r e + (b - \hat{b})u \\
&= \tilde{a} p + \tilde{b} u + a_r e.
\end{aligned} \tag{5.7}$$

A positive definite Lyapunov candidate function is selected as follow

$$V = \frac{1}{2} e^2 + \frac{1}{2}\tilde{a} + \frac{1}{2}\tilde{b}. \tag{5.8}$$

The differentiation of V is defined as

$$\dot{V} = e\dot{e} + \tilde{a}\dot{\tilde{a}} + \tilde{b}\dot{\tilde{b}}. \tag{5.9}$$

Substituting the error dynamics in above equation and we obtain

$$\dot{V} = a_r e^2 + \tilde{a}(pe - \dot{\hat{a}}) + \tilde{b}(ue - \dot{\hat{b}}). \tag{5.10}$$

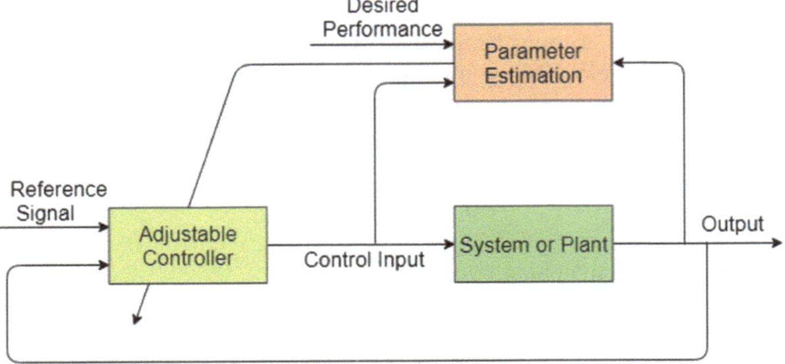

Fig. 5.3 Basic configuration of direct adaptive control

The adaptive control laws are selected as

$$\dot{\hat{a}} = p\,e \ , \quad \dot{\hat{b}} = u\,e.$$

(5.11)

Substituting the above equation in (5.10) and we obtain

$$\dot{V} = a_r e^2 \le 0,$$

(5.12)

and consequently, the other signals $(e, \tilde{a}, \tilde{b}$ are uniformly bounded. Since reference input signal, $r(t)$ is bounded and $a_r < 0$. Also, reference state p_r is also uniformly bounded and so the state p and the estimated parameters are uniformly bounded.

To prevent the controller from blowing up through division by $\hat{b}$, the adaptive law is modified as

$$\dot{\hat{b}} = \begin{cases} \gamma_b ue, & \text{if } |\hat{b}| \ge b \ \& \ uesgn(b) \ge 0 \\ 0, & \text{Otherwise}, \end{cases}$$

thereby enforcing the boundedness of the $\hat{b}$.

5.3 Direct Adaptive Control

A direct (implicit) adaptive control of chemical process is discussed in [8]. In this control scheme, on-line system model estimation is not required, instead we need to parametrize the plant model as per parameters in the controller, which are directly estimated on-line sans computations like in the explicit adaptive control [4], as described in Fig. 5.3.

Example of Direct Adaptive Control: Consider a scalar system

$$\dot{p} = ap + bu \tag{5.13}$$

where p and u refer to system state and input variable. However, a and b are known and $b \neq 0$ for controllability. The control goal is to have the actual system asymptotically behave like the chosen reference system given as

$$\dot{p}_r = a_r p_r + b_r r, \tag{5.14}$$

where $p_r \in L_\infty$ and $r(t) \in L_\infty$ refer to reference system state and reference signal respectively. Note that, a_r must be negative. The reference model is selected so as to provide desired closed-loop command tracking. The tracking error is provided by

$$e(t) = p(t) - p_r(t).$$

The differentiation of error signal is defined by

$$\dot{e}(t) = ae + bu - \dot{p}_r(t) + ap_r. \tag{5.15}$$

Assuming that the nominal parameters a and b are known, the control input signal is defined by

$$\begin{aligned} u &= \frac{1}{b}[(-a + a_r)e - a\,p_r + \dot{p}_r] \\ &= \hat{k}_1 e + \hat{k}_2 p_r + \hat{k}_3 \dot{p}_r. \end{aligned} \tag{5.16}$$

Considering that the matching conditions

$$bk_1 = -a + a_r, \quad bk_2 = -a \;, \quad bk_3 = 1, \tag{5.17}$$

are satisfied, and substituting (5.16) and (5.17) into (5.15), we obtain

$$\dot{e}(t) = a_r e - b\tilde{k}_1 e - b\tilde{k}_2 p_r - b\tilde{k}_3 \dot{p}_r \tag{5.18}$$

where $\tilde{k}_i = k_i - \hat{k}_i$, $i = 1, 2, 3$ refer to gain errors. A positive definite Lyapunov candidate function is selected as

$$V = \frac{1}{2}e^2 + \frac{|b|}{2\gamma_1}\tilde{k}_1^2 + \frac{|b|}{2\gamma_2}\tilde{k}_2^2 + \frac{|b|}{2\gamma_3}\tilde{k}_3^2. \tag{5.19}$$

The differentiation of V is defined as

$$\dot{V} = e\dot{e} - \frac{|b|}{\gamma_1}\tilde{k}_1\dot{\hat{k}}_1 - \frac{|b|}{\gamma_2}\tilde{k}_2\dot{\hat{k}}_2 - \frac{|b|}{\gamma_3}\tilde{k}_3\dot{\hat{k}}_3$$

$$= a_r e^2 - \tilde{k}_1\left(be^2 + \frac{|b|}{\gamma_1}\dot{\hat{k}}_1\right) - \tilde{k}_2\left(bep_r + \frac{|b|}{\gamma_2}\dot{\hat{k}}_2\right) - \tilde{k}_3\left(be\dot{p}_r + \frac{|b|}{\gamma_3}\dot{\hat{k}}_3\right) \quad (5.20)$$

The adaptive control laws are selected as

$$\dot{\hat{k}}_1 = -\gamma_1 e^2 sgn(b), \quad \dot{\hat{k}}_2 = -\gamma_2 ep_r sgn(b),$$
$$\dot{\hat{k}}_3 = -\gamma_3 e\dot{p}_r sgn(b). \quad (5.21)$$

Substituting (5.21) into (5.20) and we obtain

$$\dot{V} = a_m e^2 \leq 0. \quad (5.22)$$

From the above adaptive laws and the proposed control law u, we can get the desired performance by ensuring that $e \to 0$ and also ensuring $\dot{V} \leq 0$. Similar scheme may be used for other nonlinear problems where the choice of control law is known and parameters are uncertain.

5.4   Model Reference Adaptive Control

In the Model reference adaptive control (MRAC) method, a reference model with a desired input signal is needed to obtain a desired system performance. The adaptive control technique is developed as a stabilizing or a tracking controller, wherein the tracking error (the difference between the measured system output and those of the reference system) is adapted. In MRAC, a good knowledge about attributes of the reference model is utilized, such that the desired closed-loop system attributes are achieved. By choosing a proper reference model, one can develop efficient adaptive control mechanism. Typically, a reference model is an LTI model but it can also be nonlinear. A nonlinear reference model design considers complex issues [5]. A basic configuration of MRAC scheme is described in Fig. 5.4.

The adaptive control objective is to maintain a minimal tracking error by adopting proper adaptation and estimation mechanisms on the uncertain parameters. Adaptive laws are ordinary differential equations that allow adjustment of adaptive parameters to achieve this objective. Stability of the adaptive control mechanism is mathematically analyzed by Lyapunov stability theory. Number of adaptive laws depend on the number of uncertain or unknown parameters which are estimated on-line. Let us consider an example to describe the MRAC.

Example: A nonlinear plant is given as

$$\dot{x} = pf_1(x) + qf_2(x)u, \quad (5.23)$$

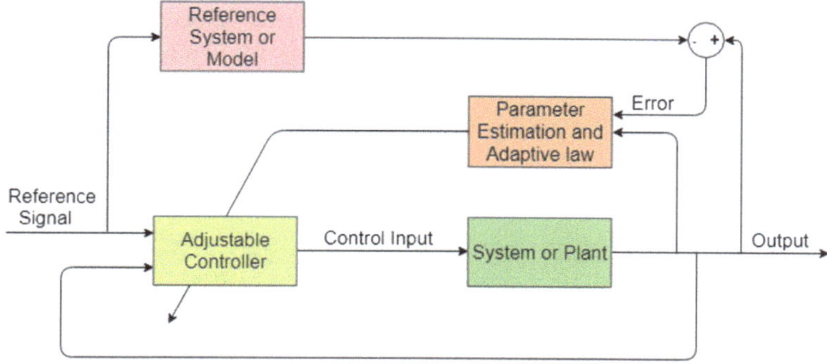

Fig. 5.4 Schematic diagram of model reference adaptive control scheme

where p, q are unknown scalars, $f_1(x)$, $f_2(x)$ are known functions and $f_1(x)$ is bounded for bounded x. The control goal of **MRAC** scheme is for the state x to track the reference state x_m, given by

$$\dot{x}_m = -p_m x_m + q_m r, \tag{5.24}$$

where r is the bounded reference signal. The control law is

$$u = \frac{1}{f_2(x)}[k_1 f_1(x) + k_2 x + lr], \quad f_2(x) \neq 0, \tag{5.25}$$

where k_1, k_2 and l are gains of the controller. If p, q are unknown, we propose a control law with adjustable gains, given by

$$u = \frac{1}{f_2(x)}[\hat{k}_1 f_1(x) + \hat{k}_2 x + \hat{l}r], \tag{5.26}$$

where $\hat{k}_1$, $\hat{k}_2$ and $\hat{l}$ are the estimates of the unknown controller gains k_1, k_2 and l respectively, and are generated by adaptive update laws. Substituting (5.25) in (5.23), we get

$$\dot{x} = pf_1(x) + q[k_1 f_1(x) + k_2 x + lr]. \tag{5.27}$$

If we choose k_1, k_2 and l as below, we could exactly meet the control objective:

$$k_1 = -\frac{p}{q}, \quad k_2 = -\frac{p_m}{q}, \quad l = \frac{q_m}{q}. \tag{5.28}$$

Substituting (5.28) in (5.27) and adding and subtracting $b(k_1 f_1(x) + k_2 x + lr)$, we obtain the dynamics

$$\dot{x} = -p_m x + q_m r + q[-k_1 f_1(x) - k_2 x - lr + f_2(x)u]. \tag{5.29}$$

Now, we define tracking error, $e = x - x_m$ and parameter errors, $\tilde{k}_1 = k_1 - \hat{k}_1, \tilde{k}_2 = k_2 - \hat{k}_2$ and $\tilde{l} = l - \hat{l}$. The derivative of tracking error is given a

$$\dot{e} = -p_m e + q[\tilde{k}_1 f_1(x) + \tilde{k}_2 x + \tilde{l} r]. \tag{5.30}$$

We select a positive definite Lyapunov candidate function as

$$V = \frac{e^2}{2} + \frac{\tilde{k}_1^2}{2\gamma_1}|q| + \frac{\tilde{k}_2^2}{2\gamma_2}|q| + \frac{\tilde{l}^2}{2\gamma_3}|q|. \tag{5.31}$$

The differentiation of V is given as

$$
\begin{aligned}
\dot{V} &= e\dot{e} + \frac{\tilde{k}_1 \dot{\hat{k}}_1}{\gamma_1}|q| + \frac{\tilde{k}_2 \dot{\hat{k}}_2}{\gamma_2}|q| + \frac{\tilde{l} \dot{\hat{l}}}{\gamma_3}|q| \\
&= -p_m e^2 + q\, e[\tilde{k}_1 f_1(x) + \tilde{k}_2 x + \tilde{l} r] + \frac{\tilde{k}_1 \dot{\hat{k}}_1}{\gamma_1}|q| + \frac{\tilde{k}_2 \dot{\hat{k}}_2}{\gamma_2}|q| + \frac{\tilde{l} \dot{\hat{l}}}{\gamma_3}|q| \\
&= -p_m e^2 + \tilde{k}_1 \left[q\, e\, f_1(x) + \frac{\dot{\hat{k}}_1}{\gamma_1}|q| \right] + \tilde{k}_2 \left[qex + \frac{\dot{\hat{k}}_2}{\gamma_2}|q| \right] \\
&\quad + \tilde{l} \left[q\, e\, r + \frac{\dot{\hat{l}}}{\gamma_3}|q| \right].
\end{aligned}
\tag{5.32}
$$

To ensure uniform system stability, $\dot{V} \leq 0$ must be satisfied, and for this purpose the adaptive laws are chosen as follows:

$$
\begin{aligned}
\dot{\hat{k}}_1 &= -\gamma_1\, e\, f_1(x)\, sgn(b) \\
\dot{\hat{k}}_2 &= -\gamma_2\, e\, x\, sgn(b), \quad \dot{\hat{l}} = -\gamma_3\, e\, r\, sgn(b).
\end{aligned}
\tag{5.33}
$$

From the above adaptive laws and the proposed control law u, we can get the desired performance by ensuring that $e \to 0$. Similar scheme may be used for other nonlinear problems where the choice of control law is known and parameters are uncertain.

5.5 Adaptive Backstepping Control

Backstepping is a Lyapunov based recursive nonlinear control technique. It is more applicable to strict feedback systems of form such as

$$\dot{p}_1 = f_1(p_1) + g_1(p_1)p_2$$
$$\dot{p}_2 = f_2(p_1, p_2) + g_2(p_1, p_2)p_3$$
$$\vdots$$
$$\dot{p}_n = f_n(p_1, p_2, ..., p_n) + g_n(p_1, p_2, ..., p_n)u,$$

where $p_1, p_2, ..., p_n \in R$ refer to the system states, control input is $u \in R$ and f_i, g_i are functions of $p_1, p_2, ..., p_n$, for $i = 1,..., n$.

The backstepping technique is considered as an alternative to feedback linearization method. It does not require the linear dynamics of the system and non-linearity cancellation and so it is able provide more flexibility compared to feedback linearization. The objective of backstepping technique is to define 'virtual control' by using some system state variables and develop intermediate control laws based on the state dynamics. The backstepping technique ensures stability and asymptotic tracking. Lyapunov function is derived for ensuring the global system stability. The adaptive controller provides online unknown parameter estimation, control law and guarantees closed-loop signal boundedness and asymptotic tracking with parametric uncertainties [6]. The design of backstepping and adaptive backstepping technique is described using following examples.

Example: Backstepping Control

Consider a strict feedback system

$$\dot{x}_1 = x_1^2 + \theta x_2 \tag{5.34}$$
$$\dot{x}_2 = u, \tag{5.35}$$

where x_1, x_2 are the system states, u is the input, and θ is the system parameter. The control goal is to obtain tracking of x_{1d} by x_1 and global system stability.

Assume that the desired or reference signal x_{1d} is constant and its derivative is zero. Define error e_1 and derive the error dynamics in terms of new coordinates

$$e_1 = x_1 - x_{1d} \tag{5.36}$$
$$\dot{e}_1 = \dot{x}_1 \;\; = \;\; x_1^2 + \theta x_2. \tag{5.37}$$

Let the error between the actual and a virtual control α be defined as

$$e_2 = x_2 - \alpha. \tag{5.38}$$

We can rewrite (5.37) as
$$\dot{e}_1 = x_1^2 + \theta(\alpha + e_2). \tag{5.39}$$

The goal is to develop virtual control law which produces error e_1 zero asymptotically. We define a Lyapunov candidate function

$$V_1 = \frac{1}{2}\theta^{-1}e_1^2. \tag{5.40}$$

The time derivative of V_1 becomes

$$\dot{V}_1 = \theta^{-1}e_1\dot{e}_1 = \theta^{-1}e_1[x_1^2 + \theta(\alpha + e_2)] = e_1[\theta^{-1}x_1^2 + \alpha + e_2]. \tag{5.41}$$

Choose a virtual control signal α given by

$$\alpha = -c_1e_1 - \theta^{-1}x_1^2, \tag{5.42}$$

where $c_1 > 0$. Substituting (5.42) in (5.41), the time derivative of V_1 becomes

$$\dot{V}_1 = e_1e_2 - c_1e_1^2. \tag{5.43}$$

Substituting (5.42) in (5.38), the error e_2 becomes

$$e_2 = x_2 + c_1e_1 + \theta^{-1}x_1^2. \tag{5.44}$$

The differentiation of error e_2 is

$$\begin{aligned}
\dot{e}_2 &= \dot{x}_2 + c_1\dot{e}_1 + \theta^{-1}2x_1\dot{x}_1 \\
&= u + (c_1 + 2\theta^{-1}x_1)(x_1^2 + \theta x_2).
\end{aligned} \tag{5.45}$$

The objective is to develop an actual control input u while ensuring that errors e_1 and e_2 approach zero over time. Define Lyapunov function V_2 as

$$V_2 = V_1 + \frac{1}{2}e_2^2. \tag{5.46}$$

The differentiation of V_2 is given by

$$\begin{aligned}
\dot{V}_2 &= \dot{V}_1 + e_2\dot{e}_2 \\
&= e_1e_2 - c_1e_1^2 + e_2[u + (c_1 + 2\theta^{-1}x_1)(x_1^2 + \theta x_2)].
\end{aligned} \tag{5.47}$$

Next, we design actual control input u so as to ensure that by $\dot{V}_2 \leq 0$ such as

$$u = -c_2e_2 - e_1 - (c_1 + 2\theta^{-1}x_1)(x_1^2 + \theta x_2), \tag{5.48}$$

where c_2 is the positive constant, Substituting (5.48) in (5.47), the time derivative of V_2 becomes

$$\dot{V}_2 = -c_1e_1^2 - c_2e_2^2 \leq 0. \tag{5.49}$$

That is, the universal system stability is confirmed. Through the above example, the design of backstepping control technique is illustrated.

Example: Adaptive Backstepping Control

Consider the system presented in the previous example, but additionally also allow the system parameters to be uncertain. The objective is the same as the previous example. That is, we are required to achieve tracking of x_{1d} by x_1 and universal system stability. Let us consider θ^{-1} as θ_1. The virtual control law α is written as

$$\alpha = -\hat{\theta}_1 x_1^2 - c_1 e_1, \tag{5.50}$$

where $\hat{\theta}_1$ is the estimation of parameter θ_1. Define a Lyapunov candidate function

$$V_1 = \frac{1}{2}\theta_1 e_1^2 + \frac{1}{2\gamma_1}\tilde{\theta}_1^2, \tag{5.51}$$

where $\tilde{\theta}_1 = \hat{\theta}_1 - \theta_1$. The derivative of Lyapunov candidate function is

$$\begin{aligned}
\dot{V}_1 &= e_1[\theta_1 x_1^2 + \alpha_1 + e_2] + \frac{1}{\gamma_1}\tilde{\theta}_1\dot{\hat{\theta}}_1 \\
&= e_1 e_2 - c_1 e_1^2 + \tilde{\theta}_1\left[e_1 x_1^2 + \frac{1}{\gamma_1}\dot{\hat{\theta}}_1\right].
\end{aligned} \tag{5.52}$$

The error e_2 with (5.50) becomes

$$e_2 = x_2 + c_1 e_1 + \hat{\theta}_1 x_1^2. \tag{5.53}$$

The derivative of error e_2 is

$$\dot{e}_2 = \dot{x}_2 + \dot{\hat{\theta}}_1 x_1^2 + 2x_1\hat{\theta}_1\dot{x}_1.$$

Substituting $\dot{x}_1$ and $\dot{x}_2$ above, we get

$$\dot{e}_2 = \dot{\hat{\theta}}_1 x_1^2 + (2x_1\theta_1 + c_1)(x_1^2 + \theta x_2) + u. \tag{5.54}$$

The objective is to develop the input u in such a way that the error signals e_1 and e_2 are guaranteed to be convergent to zero. Define a positive definite Lyapunov function V_2 as

$$V_2 = V_1 + \frac{1}{2}e_2^2 + \frac{1}{2\gamma_2}\tilde{\theta}_2^2. \tag{5.55}$$

The differentiation of V_2 is given by

$$\dot{V}_2 = \dot{V}_1 + e_2\dot{e}_2 + \frac{1}{\gamma_2}\tilde{\theta}_2\dot{\hat{\theta}}_2$$

$$= e_1e_2 - c_1e_1^2 + \tilde{\theta}_1\left[e_1x_1^2 + \frac{1}{\gamma_1}\dot{\hat{\theta}}_1\right]$$

$$+ e_2\left[\dot{\hat{\theta}}_1x_1^2 + (2x_1\theta_1 + c_1)(x_1^2 + \theta x_2) + u\right] + \frac{1}{\gamma_2}\tilde{\theta}_2\dot{\hat{\theta}}_2.$$

The control input signal is formulated by

$$u = -[\dot{\hat{\theta}}_1 + (2x_1\hat{\theta}_1 + c_1)(x_1^2 + \theta x_2) + c_2e_2 + e_1]. \tag{5.56}$$

Parameter update laws $\hat{\theta}_1$ and $\hat{\theta}$ to ensure $\dot{V}_2 \leq 0$ as follows:

$$\dot{\hat{\theta}} = \gamma_1e_1x_1^2 \tag{5.57}$$

$$\dot{\hat{\theta}}_1 = \gamma_2(2x_1\hat{\theta}_1 + c_1)x_2e_2, \tag{5.58}$$

where $c_2 > 0$. Substituting the above equations in (5.56), $\dot{V}_2$ becomes

$$\dot{V}_2 = -c_1e_1^2 - c_2e_2^2 \leq 0, \tag{5.59}$$

thereby establishing uniform stability and asymptotic tracking. The controller designed in these two examples are achieved the stabilization and tracking goals. The adaptive laws adequately estimates the uncertain parameter.

This chapter dealt with the fundamentals of adaptive control and its importance. The first part of the chapter discusses the requirement of the adaptive control techniques under uncertain parameters. A comparative study is presented between direct and indirect adaptive control. One can choose either direct or indirect adaptive control based on the dynamics and system performance requirement. The second part deals with the MRAC and the adaptive backstepping control. Suitable examples are presented to provide overview of the aforementioned adaptive techniques. One can develop these schemes as per their system dynamics.

References

1. Cao, C., Ma, L., Xu, Y.: Adaptive control theory and applications (editorial). J. Control Sci. Eng. **2012**, 1–2 (2012)
2. Feng, G., Lozano, R.: Adaptive Control Systems. Newnes (1999)
3. Ioannou, P., Sun, J.: Robust Adaptive Controls. Dover Publications Inc, New York (2012)
4. Landau, I., Lozano, R., M'Saad, M.: (1998) Adaptive Control. Springer, London (1998)
5. Nguyen, N.: Model-Reference Adaptive Control. Springer (2018)
6. Zhou, J., Wen, C.: Adaptive Backstepping Control of Uncertain Systems. Springer, Berlin (2008)

7. Chen, C.T.: Direct adaptive control of chemical process systems. Ind. Eng. Chem. Res. **40**(19), 4121–4140 (2001)
8. Selisteanu, D., Petre, E., Marin, C., Sendrescu, D.: Estimation and adaptive control of a fed-batch bioprocess. In: 2008 International Conference on Control, Automation and Systems (2008)
9. Petre, E., Selişteanu D, Şendrescu D.: Adaptive and robust-adaptive control strategies for anaerobic wastewater treatment bioprocesses. Chem. Eng. J. **217**, 363–378 (2013)
10. Dimitrova, N.S., Krastanov, M.I.: Nonlinear adaptive control of a bioprocess model with unknown kinetics. In: Modeling, Design, and Simulation of Systems with Uncertainties, pp. 275–291 (2011)
11. Selişteanu, D., Petre, E., Răsvan, V.B.: Sliding mode and adaptive sliding-mode control of a class of nonlinear bioprocesses. Int. J. Adapt. Control Signal Process. **21**(8), 795–822 (2007)

Chapter 6
Adaptive Control of Single Population Single Chamber MFC

In this chapter, a novel adaptive backstepping controller is formulated for a SPSC MFC. Adaptive control techniques are useful whenever uncertainties are present in parameters and such techniques also have an ability to estimate parameters on-line. The performance of backstepping controllers with and without parameter adaptation are analyzed and validated through appropriate simulation works.

6.1 Introduction

Attributes like relative volume of biomass and substrate, enhancement rate of microbes, operating temperature and pH in MFCs, determine system performance. Optimal performance of MFCs under different load conditions is obtained through advanced control, and the flow of MFC operation is provided in Fig. 6.1.

The output voltage available from such systems, can also vary according to environmental and operating conditions, and for stable operating voltage, controlled conditions are needed. Control oriented mathematical models of MFCs are described in the previous Chapters, which facilitate the formulation of different advanced control schemes. However, input-output linearization techniques cannot adequately represent actual MFC dynamics over a wide operating range.

Nonlinear control is widely used when the system has certain uncertainties and requires on-line parameter estimation. Advanced nonlinear control schemes provide system robustness against uncertainties. A brief summary of the various control techniques developed for different MFCs is given in Table 6.1. Voltage is considered the final output in all MFC units.

© Springer Nature Switzerland AG 2020
R. Patel et al., *Adaptive and Intelligent Control of Microbial Fuel Cells*, Intelligent Systems Reference Library 161,
https://doi.org/10.1007/978-3-030-18068-3_6

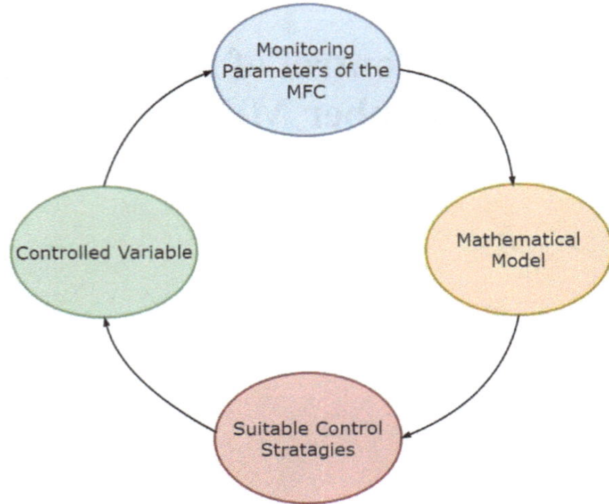

Fig. 6.1 Flow of the MFC operation with control system

Table 6.1 Control techniques developed for MFCs [1]

Compartment	Controlled input	Control techniques	Disturbance	References
Two	Flow rate	PID variants	Resistance	[2]
	Flow rate	MPC	Influent concentration and temperature	[3]
	Electrical load	Gain scheduling	Temperature and loading rate	[5]
Single (Staged)	Flow rate	On/Off and PID	Influent concentration	[4]
Two (Stacked)	Electrical load	Sampled-time digital control	Substrate concentration, temperature loading	[6]

6.2 Backstepping Control Scheme

Backstepping is a nonlinear control technique used in lieu of feedback linearization. A key advantage is the capability to control systems with relative degree greater than one, and also systems with highly nonlinear behavior. This scheme has seen increased interest due to a structured recursive approach that guarantee stability, and avoid dynamic nonlinearity cancellation. All but one states act as 'virtual controls' and intermediate control laws are formulated for such states. The backstepping approach provides provision for stabilization and tracking.

Drawbacks include the need to precisely know the system parameters and also increased complexity due to the derivative terms of virtual control signals. However, this method readily applies to nonlinear dynamical systems that are provided in affine or strict feedback forms [7]. For a SC MFC, the dynamical systems are in strict feedback form. Backstepping control is applied to various applications in literature [8–13].

MFCs can treat wastewater with a high concentration of ammonia and facilitates nitrogen (NH_4^+) removal, while acetate (substrate) helps achieve bacterial growth [14]. The dynamics of SC MFC is represented by

$$\dot{x}_1 = -\theta_1^{-1}.Y^{-1}\frac{x_1}{K_s + x_1}x_2 + u(C_{so} - x_1), \tag{6.1}$$

$$\dot{x}_2 = \left(\theta_1^{-1}\frac{x_1}{K_s + x_1} - K_d - u\right)x_2, \tag{6.2}$$

where $u(t)$ is generated through backstepping control with substrates like acetate.

The control objective is to formulate a backstepping structure for an optimal MFC performance while ensuring that the substrate concentration steadies at a desired value. We change the state variables

$$x_1 = C_{so} - z_1z_2, \quad x_2 = z_1, \tag{6.3}$$

in (6.1) and (6.2), to get

$$z_1\dot{z}_2 = \theta_1^{-1}k_1\, r(C_{so} - z_1z_2)z_1 - u\,z_1z_2 - z_2\dot{z}_1, \tag{6.4}$$

$$\dot{z}_1 = \left(\theta_1^{-1}r(C_{so} - z_1z_2) - k_d - u\right)z_1, \tag{6.5}$$

and then replace $\dot{z}_1$ in (6.4) by (6.5), to get

$$\dot{z}_2 = k_dz_2 + \theta_1^{-1}(k_1 - z_2)r(C_{so} - z_1z_2). \tag{6.6}$$

Next, we formulate error terms e_1 and e_2 as

$$e_1 = z_2 - z_2^*, \tag{6.7}$$
$$e_2 = (k_1 - z_2)r(C_{so} - z_1z_2) - \alpha_1, \tag{6.8}$$

where z_2^* and α_1 are the desired values of substrate concentration and control signal respectively. Choosing a Lyapunov candidate expression as

$$V_1 = \theta_1\frac{e_1^2}{2}, \tag{6.9}$$

and using (6.7), we get

$$\dot{V}_1 = \theta_1 e_1\dot{e}_1 = \theta_1 e_1\dot{z}_2. \tag{6.10}$$

Substituting (6.6) in (6.10), and using e_2 from (6.8), we get

$$\dot{V}_1 = \theta_1 e_1 [\theta_1^{-1}(k_1 - z_2)\, r(C_{so} - z_1\, z_2) + k_d z_2],$$
$$= e_1[e_2 + \alpha_1 + k_d z_2 \theta_1]. \tag{6.11}$$

Virtual control signal α_1 is chosen to cancel the term $k_d\, z_2\, \theta_1$ in (6.11) as

$$\alpha_1 = -C_1 e_1 - k_d z_2 \theta_1 \, , \, C_1 > 0, \tag{6.12}$$

which ensures that $\dot{V}_1$ in (6.11), is given as

$$\dot{V}_1 = e_1 e_2 - C_1 e_1^2. \tag{6.13}$$

Substituting (6.12) in (6.8), we get

$$e_2 = (k_1 - z_2) r(C_{so} - z_1 z_2) + C_1 e_1 + k_d z_2 \theta_1. \tag{6.14}$$

The first derivative of e_2 is given as

$$\dot{e}_2 = C_1 \dot{z}_2 + k_d \theta_1 \dot{z}_2 - r(C_{so} - z_1 z_2)\dot{z}_2 + (k_1 - z_2)\, \eta\, [-z_2 \dot{z}_1 - z_1 \dot{z}_2], \tag{6.15}$$

where η is defined as $\frac{k_s}{(k_s + C_{so} - z_1 z_2)^2}$. Substituting (6.5) in (6.15), we obtain

$$\dot{e}_2 = f + (k_1 - z_1)\, \eta\, z_1 z_2 u, \tag{6.16}$$

with f given as

$$f = \left[C_1 + k_d \theta_1 - r(C_{so} - z_1 z_2) - (k_1 - z_2)\, \eta\, z_1 \right] \dot{z}_2$$
$$- (k_1 - z_2)\, \eta\, z_2 \left[\theta_1^{-1} r(C_{so} - z_1 z_2) z_1 - k_d z_1 \right].$$

For cancellation of the other terms of $\dot{e}_2$, a control signal u is expressed as

$$u = \frac{1}{(k_1 - z_2)\, \eta\, z_2 z_1} \left(-C_2 e_2 - f + e_1 \right), \quad C_2 > 0. \tag{6.17}$$

The other function candidate V_2 and derivative $\dot{V}_2$ are expressed as

$$V_2 = V_1 + \frac{e_2^2}{2},$$
$$\dot{V}_2 = \dot{V}_1 + e_2 \dot{e}_2 = -C_1 e_1^2 - C_2 e_2^2. \tag{6.18}$$

The energy function V_1 is a non-increasing time function which implies that the error signals such as e_1 is bounded time function. The parameters θ_1 and θ_2 are constant and also x is bounded. Consequently, control input u is bounded. Thus, $\dot{x}_i\ i = 1, 2$

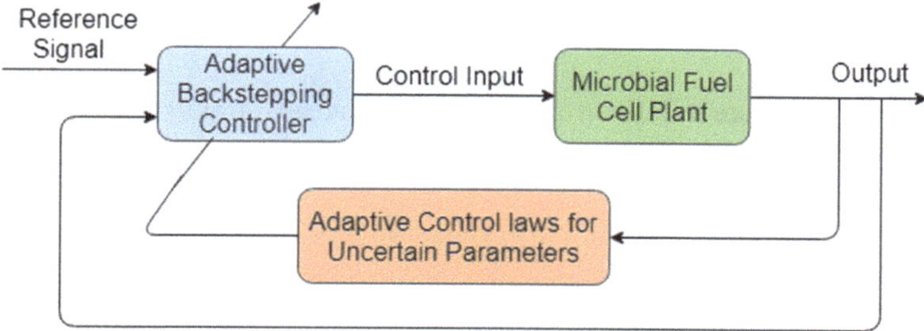

Fig. 6.2 Adaptive controller scheme for MFC

is bounded, and so $\dot{e}_i$ $i = 1, 2$ is bounded. Taking the derivative of $\dot{V}_2$

$$\ddot{V}_2 = -2c_1 e_1 \dot{e}_1 - 2c_2 e_2 \dot{e}_2,$$

where $\dot{e}_i$, $i = 1, 2$ is uniformly bounded. Thus, $\dot{V}_1$ is uniformly continuous. Also, by definition, $V_1 \geq 0$, and since $\dot{V}_1 \leq 0$, V_1 is non-increasing and tends to a limit as $t \to \infty$. We have shown that $0 \leq V_1 < \infty$ and $\dot{V}_1$ are uniformly continuous. According to the corollary of the Barbalat's criteria

$$\lim_{t \to \infty} e_1(t) = 0.$$

As a result of (6.13), (6.18) turns out to be negative definite in steady state. Hence, with virtual control α_1, we have achieved global asymptotic stability [19].

6.3 Adaptive Backstepping Control Scheme

The adaptive version of backstepping control designs is an advanced research direction of control theory, that offers a systematic control approach for a diverse class of nonlinear systems with known structure but uncertain parameters. They also facilitate an improved closed-loop system transient behavior even in absence of uncertainties. The process and model outputs may be compared and the error function minimized via appropriate optimization routines so as to achieve re-tuned controller parameters in real-time.

An adaptive backstepping controller is presented in case of a SC MFC for improved response to unmodeled and parametric uncertainties. An online parameter estimation for the parameter θ_1 is performed. A block diagram with adaptive update laws for estimated parameters and control action is given in Fig. 6.2.

6.3.1 Adaptive Controller Design

The parameter estimate and the error estimate terms are given as $\hat{\theta}_1$ and $\tilde{\theta}_1 (= \hat{\theta}_1 - \theta_1)$ respectively. The virtual control signal is expressed as

$$\alpha_1 = -C_1 e_1 - k_d z_2 \hat{\theta}_1. \tag{6.19}$$

The Lyapunov candidate function and the corresponding time derivative expression are defined as

$$V_{s_1} = V_1 + \frac{\tilde{\theta}_1^2}{2\gamma_1}, \quad \dot{V}_{s_1} = \dot{V}_1 + \frac{\tilde{\theta}_1 \dot{\hat{\theta}}_1}{\gamma_1}, \quad \gamma_1 > 0. \tag{6.20}$$

Substituting for α_1, from (6.19) in (6.11), we get

$$\dot{V}_1 = e_1 e_2 - c_1 e_1^2 - k_d z_2 e_1 \tilde{\theta}_1, \tag{6.21}$$

which on substitution in (6.20), results in

$$\dot{V}_{s_1} = e_1 e_2 - C_1 e_1^2 + \tilde{\theta}_1 \left[-k_d z_2 e_1 + \frac{\dot{\hat{\theta}}_1}{\gamma_1} \right]. \tag{6.22}$$

Using (6.8) and (6.19), e_2 is given as

$$e_2 = (k_1 - z_2) r (C_{so} - z_1 z_2) + C_1 e_1 + k_d z_2 \hat{\theta}_1. \tag{6.23}$$

The derivative of e_2 is

$$\dot{e}_2 = p \dot{z}_2 - (k_1 - z_2) \eta z_2 \dot{z}_1 + k_d z_2 \dot{\hat{\theta}}_1, \tag{6.24}$$

where $p = C_1 + k_d \hat{\theta}_1 - r(C_{so} - z_1 z_2) - (k_1 - z_2)\eta z_1$. Using (6.5) and (6.6), we get

$$\dot{e}_2 = \theta_2 h(z_1, z_2, \hat{\theta}_1) + f(z_1, z_2, \hat{\theta}_1) + k_d z_2 \dot{\hat{\theta}}_1 + (k_1 - z_2)\eta z_1 z_2 u, \tag{6.25}$$

where

$$h(z_1, z_2, \hat{\theta}_1) = (k_1 - z_2)(p - \eta z_2 z_1) r (C_{so} - z_1 z_2),$$
$$f(z_1, z_2, \hat{\theta}_1) = (k_1 - z_2)\eta z_1 z_2 k_d + p k_d z_2,$$
$$\theta_2 = \theta_1^{-1}.$$

A suitable adaptive update law is expressed as

$$u = \frac{-1}{(k_1 - z_2)z_1 z_2 \eta}\left[C_2 e_2 + e_1 + k_d z_2 \dot{\hat{\theta}}_1 + f(z_1, z_2, \hat{\theta}_1) + \hat{\theta}_2 h(z_1, z_2, \hat{\theta}_1)\right]. \quad (6.26)$$

Substituting u in (6.25), we obtain

$$\dot{e}_2 = -C_2 e_2 - \tilde{\theta}_2 h(z_1, z_2, \hat{\theta}_1) - e_1. \quad (6.27)$$

6.3.2 Adaptive Update Laws

The adaptive update laws guarantee optimality of the control laws for the closed loop dynamics. We design a novel parameter projection based backstepping controller with adaptive update of control parameters θ_i, $i = 1, 2$ for signal boundedness ($\theta_{i,min} \leq \hat{\theta}_i \leq \theta_{i,max}$, such that the bounds are known and $\theta_{i,min}, \theta_{i,max} > 0$). It is also assumed that initial conditions for parameter estimates are such that $\theta_{i,min} \leq \hat{\theta}_{i,0} \leq \theta_{i,max}$ is guaranteed. The adaptive update laws for θ_1 and θ_2 are

$$\dot{\hat{\theta}}_i = g_i + f_i, \quad i = 1, 2, \quad t \geq 0, \quad (6.28)$$

where $g_1 = k_d z_2 e_1 \gamma_1$, $g_2 = e_2 h(z_1, z_2, \hat{\theta}_1)\gamma_2$, and adaption gains $\gamma_i > 0$, $i = 1, 2$ and $h(z_1, z_2, \hat{\theta}_1)$ are obtained from Sect. 3.2.1, and f_i is defined as

$$f_i = \begin{cases} 0 & \text{if } \theta_{i,min} \leq \hat{\theta}_i \leq \theta_{i,max} \\ -g_i & \text{otherwise.} \end{cases}$$

6.3.3 Stability Performance Analysis

The proposed adaptive controller structure guarantees asymptotic system stability. Tracking error is defined as the error between the measured and the desired substrate concentration, while the parameter errors are the errors between the parameter estimates and actual parameters respectively. It is important to ensure that the tracking errors asymptotically approach zero and the parameters are tightly bounded.

Theorem *The adaptive backstepping controller with a input signal u as given in (6.26) wherein the parameter estimates are governed by update laws in (6.28), when provided to the mathematical model of MFC, provides closed-loop signal boundedness and tracking error $\lim_{t \to \infty} e(t) = 0$ [15–18].*

$\triangledown$

Proof We select a Lyapunov candidate function $V_{s_2}(e_2, \tilde{\theta}_2)$ expressed as

$$V_{s_2} = V_{s_1} + \frac{e_2^2}{2} + \frac{\tilde{\theta}_2^2}{2\gamma_2}, \quad \gamma_2 > 0. \tag{6.29}$$

The differentiation of V_{s_2} is

$$\dot{V}_{s_2} = \dot{V}_{s_1} + e_2\dot{e}_2 + \frac{\tilde{\theta}_2\dot{\hat{\theta}}_2}{\gamma_2}. \tag{6.30}$$

Substituting (6.22) and (6.25) in (6.30), we get

$$\dot{V}_{s_2} = e_1e_2 - C_1e_1^2 - C_2e_2^2 + \tilde{\theta}_1\left[-k_dz_2e_1 + \frac{\dot{\hat{\theta}}_1}{\gamma_1}\right] + \tilde{\theta}_2\left[-e_2h(z_1, z_2, \hat{\theta}_1) + \frac{\dot{\hat{\theta}}_2}{\gamma_2}\right].$$

Substituting the update laws (6.28) in (6.30), we have

$$\dot{V}_{s_2} = -C_1e_2^2 - C_2e_2^2. \tag{6.31}$$

For $\theta_{i,min} \leq \hat{\theta}_i \leq \theta_{i,max}$, $\theta_{i,min}, \theta_{i,max} > 0$, and $C_1, C_2 > 0$, we can state that

$$\dot{V}_{s_2} = -C_1e_2^2 - C_2e_2^2 \leq 0. \tag{6.32}$$

The energy function V is non-increasing which implies that all the error signals like e, $\tilde{\theta}_1$, and $\tilde{\theta}_2$ are bounded function of time. Since the θ_1 and θ_2 are constant, the estimates $\hat{\theta}_1$ and $\hat{\theta}_2$ are also bounded. Consequently, u is bounded. Thus, $\dot{x}_i$ $i = 1, 2$ are bounded, and $\dot{e}_i$ $i = 1, 2$ are also bounded. Taking the derivative of $\dot{V}_{s_2}$, we get

$$\ddot{V}_{s_2} = -2c_1e_1\dot{e}_1 - 2c_2e_2\dot{e}_2,$$

where $\dot{e}_i, i = 1, 2$ are uniformly bounded functions. Thus, $\dot{V}$ is uniformly continuous. By definition, $V \geq 0$, and since $\dot{V} \leq 0$, V is non-increasing and tends to a limit as $t \to \infty$. We have shown that $0 \leq V < \infty$ and $\dot{V}$ are uniformly continuous. As per the corollary of the Barbalat's lemma

$$\lim_{t \to \infty} e(t) = 0.$$

Since, $e_1, e_2, \theta_1, \theta_2$ and $\dot{e}_1$ are uniformly bounded, we have $\lim_{t \to \infty} e(t)=0$ [15–19]. Thus, an adaptive backstepping controller effectively ensures that the tracking error approaches zero, thereby providing satisfactory MFC performance.

The efficacy of the proposed methodology is evaluated through a realistic simulation study performed in MATLAB/ Simulink to understand the processes of utilization of the substrate, evolution of the biomass concentration, and generation of voltage across a load resistor. Nominal parameter values, constants and conditions at the start of simulation, are already given in Table 2.3. Additional constants and adaptive gains are appropriately taken as $C_1 = 0.13, C_2 = 0.4, \gamma_1 = 0.03, \gamma_2 = 50$.

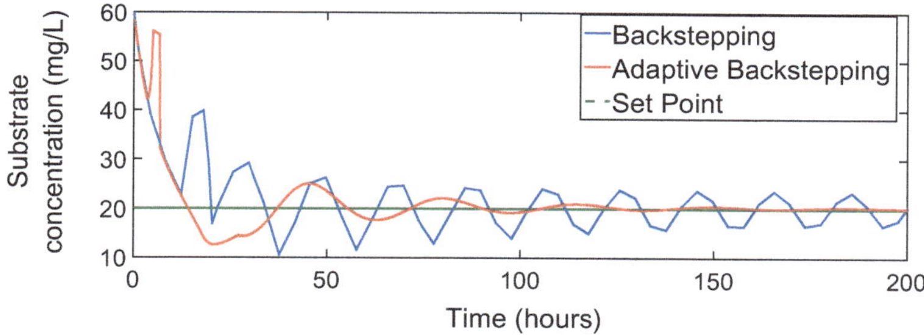

Fig. 6.3 Performance of substrate concentration (Desired)

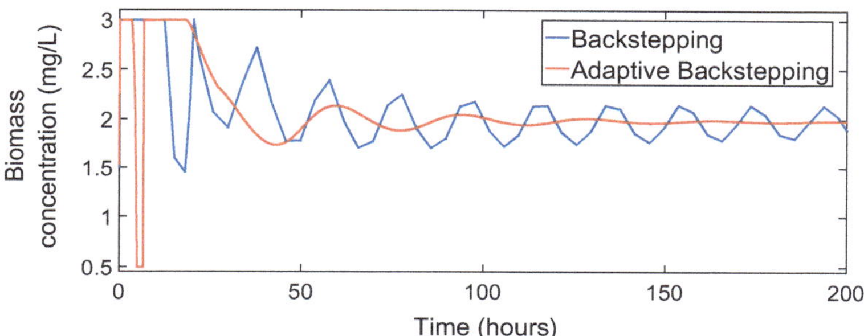

Fig. 6.4 Performance of biomass concentration

The objective is to secure a steady output while controlling the substrate at a desired concentration in presence of varied conditions of load. The desired goal is to have the tracking and parameter errors approach zero or nearer to zero at the steady state. The performance of the two control methodologies are compared to each other, as described in Fig. 6.3.

It is observed that the substrate concentration switches within a range of 18–23 mg/L and is not entirely steady at the desired set point of 20 mg/L without parameter adaptation. The adaptive phenomenon ensures that the substrate concentration is driven to a desired level. The concentration of biomass which is not the controlled variable, is limited within [0.5, 3] mg/L, to guarantee a desired outcome as in Fig. 6.4. We observe a spike at 5–10 h because the parameter estimates are bounded.

The adaptive controller us able to estimate parameters online, and is structured such that the tracking error pertaining to the actual and the desired substrate concentration, goes to zero. The nominal parameter are reasonably assumed to bounded in the following ranges: $2 < \theta_1 < 2.857$ and $0.35 < \theta_2 < 0.5$. Figures 6.5, 6.6 and 6.7 describe the traversal of the tracking error signal, parameter estimates of θ_1 and θ_2 with adaptive update, and parameters errors that approach zero at steady state.

The primary purpose of application of any sort of control techniques on MFCs is to ensure almost fixed voltage at the output under diverse conditions of loads. The traversal of the voltage values at the two electrodes, and the MFC output voltage are presented in Figs. 6.8, 6.9 and 6.10 respectively.

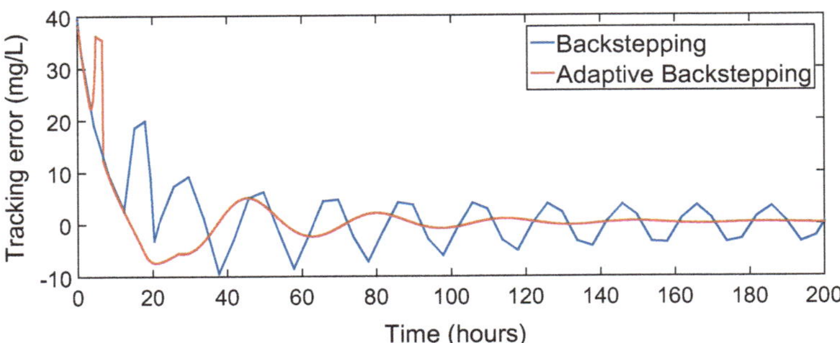

Fig. 6.5 Tracking error comparison

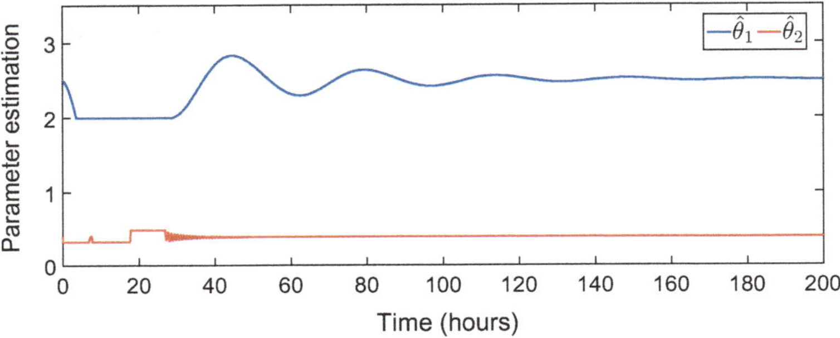

Fig. 6.6 Parameters estimations

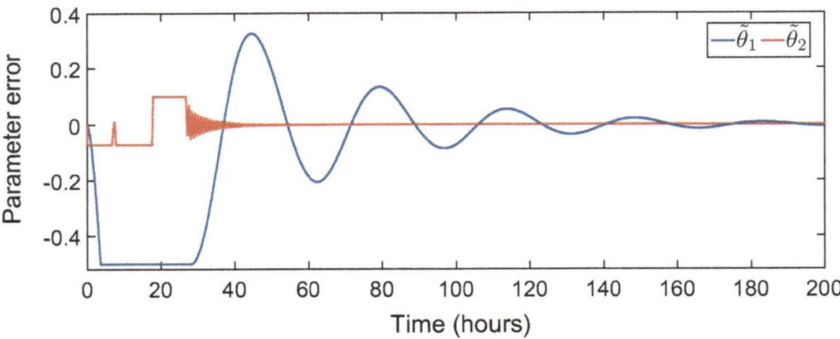

Fig. 6.7 Parameter errors

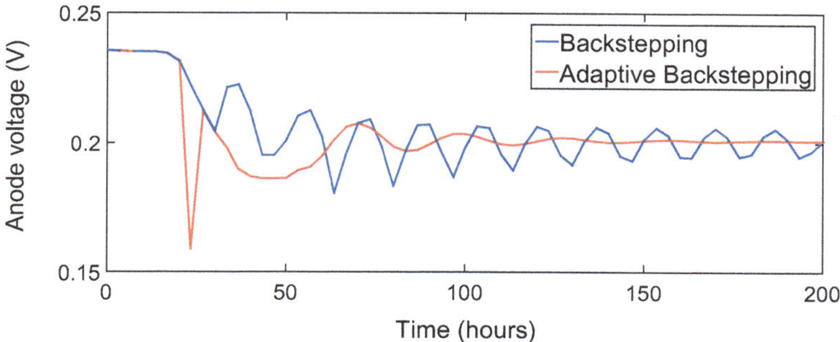

Fig. 6.8   Anode voltage

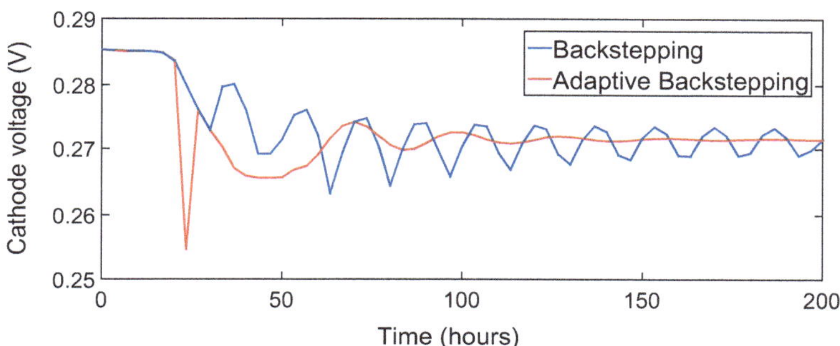

Fig. 6.9   Cathode voltage

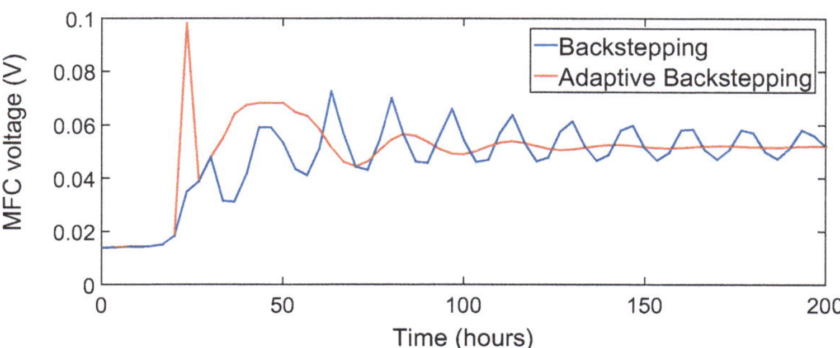

Fig. 6.10   MFC voltage

In the non-adaptive scenario, the fluctuations seen are a result of the switching of the concentration of the substrate. The voltage readings at the anode, cathode and the output voltage as determined from the Nernst equation, are provided in Table 2.1. Reaction quotient Q depends on the states. With adaptive control, we observe a pronounced increase in voltage around 20 h after the start of the simulation, because of the states and Q. With uncertainty in the parameters within the given range, adaptive backstepping controller is found to be efficient as compared to other control methodologies (Fig. 6.9).

An effective formulation of a backstepping controller with adaptive action is needed such that proper parameters are selected for a better system performance. Inappropriate choice of parameters or parameter bounds might lead to inadequate system response or even instability [7]. This methodology is more effective compared to conventional techniques in improving transient performance of adaptable systems through the process of appropriate online parameter tuning in the presence of parametric uncertainties [20].

This chapter dealt with the development of nonlinear control strategy for SPSC MFC. The first part of the chapter discussed the basics of backstepping control technique, its advantages and drawbacks, and the design of a Backstepping controller for a SPSC MFC. The second part dealt with the adaptive backstepping control scheme whose effectiveness is validated through simulation work, and performance comparison of backstepping and adaptive backstepping methods is presented, and the latter is found to provide better performance under parametric uncertainties. The stability of adaptive control scheme is evaluated through the Lyapunov second method of stability analysis. Considering the trade-off between higher generation of electrical output as a result of the attached biomass concentration and acceptable energy efficiency, it is important to determine an appropriate rate of influent concentration.

References

1. Recio-Garrido, D., Perrier, M., Tartakovsky, B.: Modeling, optimization and control of bio-electrochemical systems. Chem. Eng. J. **289**, 180–190 (2016)
2. Fan, L., Li, C., Boshnakov, K.: Performance improvement of a microbial fuel cell based on adaptive fuzzy control. Pak. J. Pharm. Sci. **10**, 685–690 (2015)
3. Fan, L., Zhang, J., Shi, X.: Performance improvement of a microbial fuel cell based on model predictive control. Pak. J. Pharm. Sci. **10**, 737–748 (2015)
4. Recio-Garrido, D., Tartakovsky, B., Perrier, M.: Staged microbial fuel cells with periodic connection of external resistance. IFAC-PapersOnLine **49**, 91–96 (2016)
5. Boghani, H.C., Michie, I., Dinsdale, R.M., Guwy, A.J., Premier, G.C.: Control of microbial fuel cell voltage using a gain scheduling control strategy. J. Power Sour. **322**, 106–115 (2016)
6. Boghani, H.C., Dinsdale, R.M., Guwy, A.J., Premier, G.C.: Sampled-time control of a microbial fuel cell stack. J. Power Sour. **356**, 338–347 (2017)
7. Ariffanan, M., Basri, M., Husain, A., Danapalasingam, K.: Backstepping Controller with Intelligent Parameters Selection for Stabilization of Quadrotor Helicopter. J. Eng. Sci. Technol. Rev. **7**(2), 66–74 (2016)

8. Ye, L., Zong, Q., Tian, B., Zhang, X., Wang, F.: Control-oriented modeling and adaptive backstepping control for a nonminimum phase hypersonic vehicle. ISA Trans. **70**, 161–172 (2017)
9. Yu, J., Shi, P., Zhao, L.: Finite-time command filtered backstepping control for a class of nonlinear systems. Automatica **92**, 173–180 (2018)
10. Zhu, B., Huo, W.: Adaptive backstepping control for a miniature autonomous helicopter. In: IEEE Conference on Decision and Control and European Control Conference (2011)
11. Nizami, T.K., Chakravarty, A., Mahanta, C.: Analysis and experimental investigation into a finite time current observer based adaptive backstepping control of buck converters. J. Frankl. Inst. **355**(12), 4996–5017 (2018)
12. Patel, R., Deb, D., Modi, H., Shah, S.: Adaptive backstepping control scheme with integral action for quanser 2-dof helicopter. In: 2017 International Conference on Advances in Computing, Communications and Informatics (ICACCI) (2017)
13. Xian, B., Guo, J., Zhang, Y.: Adaptive backstepping tracking control of a 6-DOF unmanned helicopter. IEEE/CAA J. Autom. Sin. **2**(1), 19–24 (2015). https://doi.org/10.1109/jas.2015.7032902
14. Capodaglio, A., Molognoni, D., Pons, A.: A multi-perspective review of microbial fuel-cells for wastewater treatment: Bio-electro-chemical, microbiologic and modeling aspects. Technologies and Materials for Renewable Energy, Environment and Sustainability (2016)
15. Deb, D., Tao, G., Burkholder, J., Smith, D.R.: Adaptive synthetic jet actuator compensation for a nonlinear tailless aircraft model at low angles of attack. IEEE Trans. Control Syst. Technol. **16**(5), 983–995 (2008)
16. Deb, D., Tao, G., Burkholder, J., Smith, D.R.: An adaptive inverse control scheme for a synthetic jet actuator model. In: Proceedings of 2005 Am. Control Conference, pp. 2646–2651 (2005)
17. Nath, A., Deb, D., Dey, R., Das, S.: Blood glucose regulation in type 1 diabetic patients: an adaptive parametric compensation control-based approach. IET Syst. Biol. **12**(5), 219–225 (2018)
18. Patel, R., Deb, D.: Parametrized control-oriented mathematical model and adaptive backstepping control of a single chamber single population microbial fuel cell. J. Power Sour. **396**, 599–605 (2018)
19. Lavretsky, E., Wise, K.A.: Advance Textbooks in Control and Signal Processing, Robust and Adaptive Control (2013)
20. Zhou, J., Wen, C.: Adaptive backstepping control of uncertain nonlinear systems with input quantization. In: 52nd IEEE Conference on Decision and Control (2013)

Chapter 7
Adaptive Control of Single Chamber Two-Population MFC

In this chapter, an adaptive state feedback control technique is developed for SC MFC with two bacterial species. The stability of system in closed-loop is evaluated by Lyapunov stability analysis and the system performance in presence of parametric uncertainties is validated through appropriate simulation work.

7.1 Introduction

Controllable model of SC MFC with two bacterial species is developed as follows. The dynamics of MFC are represented by (2.7)–(2.9), parameters (μ_{max} and q_{max}) and certain constants.

Let x_1 be the concentration of anodophilic microorganisms (X_a), x_2 be the concentration of methanogenic microorganisms (X_m) and x_3 be the substrate concentration (C_s). Input u represents the influent substrate concentration (C_{so}). Dilution rate (D) can also be a manipulated variable. However, while using D as an input, large control action is required which causes sluggish response of the controlled variable. Therefore influent substrate concentration (C_{so}) is used as a manipulated so as to meet the control objective [1].

The choice of the manipulated input variable is completely depend on the system performance requirement and chosen control strategy. The parametrized model is given by

$$\dot{x}_1 = \theta_1 \frac{x_3}{k_\alpha + x_3} x_1 - k_a x_1 - \alpha_a D x_1$$

$$\dot{x}_2 = \theta_2 \frac{x_3}{k_\beta + x_3} x_2 - k_d x_2 - \alpha_m D x_2$$

$$\dot{x}_3 = -k_1 \theta_1 \frac{x_3}{k_\alpha + x_3} x_1 - k_2 \theta_2 \frac{x_3}{k_\beta + x_3} x_2 + D[u(t) - x_3], \tag{7.1}$$

© Springer Nature Switzerland AG 2020
R. Patel et al., *Adaptive and Intelligent Control of Microbial Fuel Cells*, Intelligent Systems Reference Library 161,
https://doi.org/10.1007/978-3-030-18068-3_7

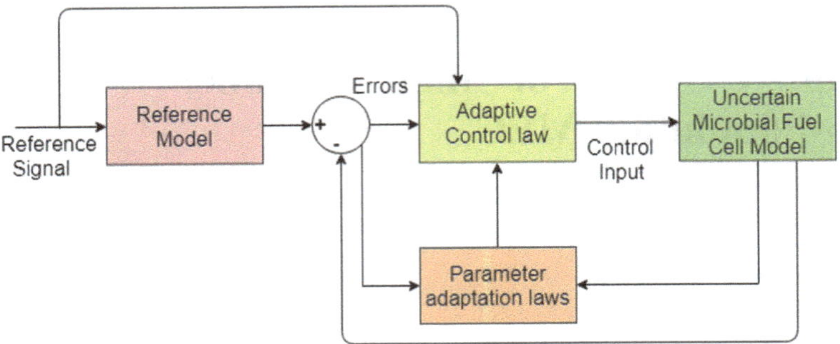

Fig. 7.1 Adaptive controller for MFC system

where parameters θ_1 and θ_2 are $\mu_{max,a}.r(M_{ox})$ and $\mu_{max,m}$ respectively, $r(M_{ox})$ is defined as $\frac{M_{ox}}{K_M+M_{ox}}$ and we assume that M_{ox} is available for measurement. Constants k_1 and k_2 are defined as the inverse of the bacterial yield Y_a and Y_m respectively. This model is highly nonlinear in nature with two bacterial species. Therefore the complexity of the kinetics in microorganisms is more in this case as compared to a single species model.

The control operation of MFCs require boundedness of the parameters. The model dynamics is based on bacterial growth and biomass consumption. The output voltage, current and power are completely depended on the state variables and parameters of the model dynamics. The relationship between state variables and output current or voltage is given in [2]. The control oriented mathematical models of different MFCs given in [3]. The similar adaptive compensation control strategy is applied to blood glucose regulation model [4].

7.2   Adaptive Control Design

A block diagram of MFC with proposed adaptive control scheme is shown in Fig. 7.1. The errors are obtained from the actual plant and reference model. Parameter adaption laws are combination of the dynamics of actual system and errors. The adaptive input control law is developed by combining of reference input signals, errors and parameter adaption laws.

A reference system of MFC is chosen with desired properties and parameters which are estimates of the uncertain system. Such a system is given by

$$\dot{\hat{x}}_1 = \hat{\theta}_1 \frac{\hat{x}_3}{k_\alpha + \hat{x}_3}\hat{x}_1 - k_a\hat{x}_1 - \alpha_a D\hat{x}_1$$

$$\dot{\hat{x}}_2 = \hat{\theta}_2 \frac{\hat{x}_3}{k_\beta + \hat{x}_3}\hat{x}_2 - k_d\hat{x}_2 - \alpha_m D\hat{x}_2$$

$$\dot{\hat{x}}_3 = -k_1\hat{\theta}_1 \frac{\hat{x}_3}{k_\alpha + \hat{x}_3}\hat{x}_1 - k_2\hat{\theta}_2 \frac{\hat{x}_3}{k_\beta + \hat{x}_3}\hat{x}_2 + D[r(t) - \hat{x}_3], \tag{7.2}$$

Where $\hat{x}_1$, $\hat{x}_2$ and $\hat{x}_3$ are the reference states, $\hat{\theta}_1$ and $\hat{\theta}_2$ are estimated parameters in the reference system and $r(t)$ is the desired reference input signal.

Reference input signal may be obtained by several experimental results which gives optimal and reasonable system performance. Known parameters and constants of the uncertain system are same in the reference model. A reference system is a known, stable, and controlled model chosen as directed by the control engineer.

Subtracting (7.1) from (7.2), the state error signals $e_i = \hat{x}_i - x_i$, $i = 1, 2, 3$, are

$$\dot{e}_1 = -k_a e_1 - \alpha_a e_1 D + \frac{1}{(k_\alpha + \hat{x}_3)(k_\alpha + x_3)}\Big[\hat{\theta}_1(k_\alpha e_3\hat{x}_1 + k_\alpha e_1 x_3 + x_3 e_1\hat{x}_3)$$
$$+\tilde{\theta}_1(k_\alpha x_1 x_3 + x_3 x_1\hat{x}_3)\Big]$$

$$\dot{e}_2 = -k_d e_2 - \alpha_m e_2 D + \frac{1}{(k_\beta + \hat{x}_3)(k_\beta + x_3)}\Big[\hat{\theta}_2(k_\beta e_3\hat{x}_2 + k_\beta e_2 x_3 + x_3 e_2\hat{x}_3)$$
$$+\tilde{\theta}_2(k_\beta x_2 x_3 + x_3 x_2\hat{x}_3)\Big]$$

$$\dot{e}_3 = \frac{1}{(k_\alpha + \hat{x}_3)(k_\alpha + x_3)}\Big[-k_1\hat{\theta}_1(k_\alpha e_3\hat{x}_1 + k_\alpha e_1 x_3 + x_3\hat{x}_3 e_1) +$$
$$\tilde{\theta}_1(k_\alpha x_1 x_3 + x_3\hat{x}_3 x_1)\Big] + \frac{1}{(k_\beta + \hat{x}_3)(k_\beta + x_3)}\Big[-k_2\hat{\theta}_2(k_\beta e_3\hat{x}_2 + k_\beta e_2 x_3$$
$$+x_3\hat{x}_3 e_2) + \tilde{\theta}_2(k_\beta x_2 x_3 + x_3\hat{x}_3 x_2)\Big] - e_3 D + [r(t) - u(t)] D, \tag{7.3}$$

where $\tilde{\theta}_i = \hat{\theta}_i - \theta_i$, $i = 1, 2$ are the parameter errors. Possible singularities in these error terms can be avoided because of the constants k_α and k_β available in (7.3).

For closed-loop stability, formulate adaptive laws to update the estimates

$$\dot{\hat{\theta}}_1 = \frac{-1}{\gamma_1}\Big[(c_1 e_1 - k_1 e_3 c_3)(k_\alpha x_1 x_3 + x_3\hat{x}_3 x_1)\Big],$$

$$\dot{\hat{\theta}}_2 = \frac{-1}{\gamma_2}\Big[(c_2 e_2 - k_2 e_3 c_3)(k_\beta x_2 x_3 + x_3\hat{x}_3 x_2)\Big]. \tag{7.4}$$

Next, for satisfactory system performance with respect to the reference system, we strategically choose an adaptive update law $u(t)$ expressed as

$$u = \frac{1}{Dc_3}\left(\frac{\hat{\theta}_1}{(k_\alpha + \hat{x}_3)(k_\alpha + x_3)}\Big[c_1 e_1 k_\alpha \hat{x}_1 - k_1 c_3(k_\alpha e_3 \hat{x}_1 + k_\alpha e_1 x_3 + x_3 e_1 \hat{x}_3)\Big] + \right.$$

$$\left.\frac{\hat{\theta}_2}{(k_\beta + \hat{x}_3)(k_\beta + x_3)}\Big[c_2 e_2 k_\beta \hat{x}_2 - k_2 c_3(k_\beta e_3 \hat{x}_2 + k_\beta e_2 x_3 + x_3 e_2 \hat{x}_3)\Big]\right) + r, \quad (7.5)$$

where c_1, c_2 and $c_3 > 0$ are constants to ensure closed loop stability using Lyapunov stability analysis.

Theorem *The adaptive control technique with input control signal $u(t)$ in (7.5) whose parameters are estimated through update laws (7.4), and applied to error dynamics (7.3), ensure that closed loop signals are bounded and the tracking errors asymptotically approach zero, that is,* $\lim_{t \to \infty} e_i(t) = 0$, $i = 1, 2, 3$ *[5, 6].*

Proof Consider a positive definite function given by

$$V = \frac{1}{2}c_1 e_1^2 + \frac{1}{2}c_2 e_2^2 + \frac{1}{2}c_3 e_3^2 + \frac{1}{2}\gamma_1 \tilde{\theta}_1^2 + \frac{1}{2}\gamma_2 \tilde{\theta}_2^2$$

as a measure of the parameter errors $\theta_i(t)$ and system errors $e_i(t)$. By differentiating the parameter errors $\tilde{\theta}_i$, we obtain

$$\dot{\tilde{\theta}}_i = \dot{\hat{\theta}}_i - \dot{\theta}_i = \dot{\hat{\theta}}_i, \quad i = 1, 2, 3. \tag{7.6}$$

Therefore, the differentiation of V is

$$\dot{V} = c_1 e_1 \dot{e}_1 + c_2 e_2 \dot{e}_2 + c_3 e_3 \dot{e}_3 + \gamma_1 \tilde{\theta}_1 \dot{\hat{\theta}}_1 + \gamma_2 \tilde{\theta}_2 \dot{\hat{\theta}}_2,$$

where $\dot{\theta}_i$ vanishes as θ_i are uncertain but constants. Substituting $\dot{e}_i$ from (7.3), the differentiation of V is

$$\dot{V} = -c_1 k_a e_1^2 - c_1 \alpha_a D e_1^2 - c_2 k_d e_2^2 - c_2 \alpha_m D e_2^2 - c_3 D e_3^2$$

$$+ \frac{1}{(k_\alpha + \hat{x}_3)(k_\alpha + x_3)}\left(\tilde{\theta}_1\Big[\dot{\hat{\theta}}_1 \gamma_1 + (c_1 e_1 - k_1 e_3 c_3)(k_\alpha x_1 x_3 + x_3 \hat{x}_3 x_1)\Big] \right.$$

$$\left. + \hat{\theta}_1\Big[(c_1 e_1 - k_1 c_3 e_3)(k_\alpha e_3 \hat{x}_1 + k_\alpha e_1 x_3 + x_3 e_1 \hat{x}_3)\Big]\right) + c_3 e_3 D r(t)$$

$$+ \frac{1}{(k_\beta + \hat{x}_3)(k_\beta + x_3)}\left(\tilde{\theta}_2\Big[(c_2 e_2 - k_2 e_3 c_3)(k_\beta x_2 x_3 + x_3 \hat{x}_3 x_2) + \dot{\hat{\theta}}_2 \gamma_2\Big] \right.$$

$$\left. + \hat{\theta}_2\Big[(c_2 e_2 - k_2 e_3 c_3)(k_\beta e_3 \hat{x}_2 + k_\beta e_2 x_3 + x_3 e_2 \hat{x}_3)\Big]\right) - c_3 e_3 D u(t). \tag{7.7}$$

Substituting the adaptive laws $\dot{\hat{\theta}}_i$ from (7.4), and control law (7.5), we get

$$\dot{V} = -c_1 e_1^2 \left(k_a + \alpha_a D - \frac{\hat{\theta}_1 x_3}{k_\alpha + x_3} \right) - c_3 e_3^2 D - c_2 e_2^2 \left(k_d + \alpha_m D - \frac{\hat{\theta}_2 x_3}{k_\beta + x_3} \right).$$

For closed loop system stability, D is chosen such that

$$D > max \left[\frac{\frac{\hat{\theta}_1 x_3}{k_\alpha + x_3} - k_a}{\alpha_a}, \frac{\frac{\hat{\theta}_2 x_3}{k_\beta + x_3} - k_d}{\alpha_m} \right]$$

and so at all times, $\dot{V} \leq 0$.

The energy function V is a non-increasing time function which implies that all the error signals like e, $\tilde{\theta}_1$, and $\tilde{\theta}_2$ are bounded functions of time. Since θ_1 and θ_2 are constants, the adaptive estimates $\hat{\theta}_1$ and $\hat{\theta}_2$ are also bounded. The design of reference system provides bounds $\hat{x}_i$ $i = 1, 2, 3$ and the reference signal $r(t)$ is bounded by design, and also the signal x is bounded. Consequently, the input signal u is also bounded. Thus, $\dot{x}$ and $\hat{x}_i$ $i = 1, 2, 3$ are also bounded. Therefore, $\dot{e}_i$ $i = 1, 2, 3$ are bounded. Taking the derivative of $\dot{V}$

$$\ddot{V} = -2c_1 e_1 \dot{e}_1 \left(k_a + \alpha_a D - \frac{\hat{\theta}_1 x_3}{k_\alpha + x_3} \right) - 2c_3 e_3 \dot{e}_3 D - 2c_2 e_2 \dot{e}_2 \left(k_d + \alpha_m D - \frac{\hat{\theta}_2 x_3}{k_\beta + x_3} \right)$$

$\dot{e}_i$, $i = 1, 2, 3$ is a uniformly bounded time function. Thus, $\ddot{V}$ is a uniformly continuous function. Since, $V \geq 0$, V is a non increasing function of time, and $\dot{V} \leq 0$, we have V tends to a limit as $t \to \infty$. We have shown that $0 \leq V < \infty$ and $\dot{V}$ are uniformly continuous. Therefore, as per Barbalat's lemma

$$\lim_{t \to \infty} e(t) = 0.$$

Thus, the adaptive controller (with update laws) forces x to asymptotically track the reference signal x_m irrespective of initial conditions. Additionally, since, e_1, e_2, e_3, $\hat{\theta}_1$ and $\hat{\theta}_2$ are uniformly bounded, and therefore $\lim_{t \to \infty} e_i(t) = 0$ [5, 7, 8].

7.3 Simulation Results

Simulation studies are undertaken to validate the effectiveness of control scheme for MFC with two species. The controller gains and constants are given in Table 7.1.

The constants c_i, $i = 1, 2, 3$ are first fixed to provide closed-loop stability using Lyapunov stability analysis. Then gains γ_1 and γ_2 are tuned to improve transient behavior and achieve the control objective of closed-loop performance and a desired steady state despite uncertainties in parameters.

Table 7.1 Controller gains and constants

Constants	Values	Gains	Values
c_1	0.01	γ_1	1.015
c_2	0.02	γ_2	2.25
c_3	0.00425		

Table 7.2 Nominal and range of parameters

Parameters	Nominal value	Range
θ_1	1.4	[0.1, 2.5]
θ_2	0.1	[0.05, 0.5]

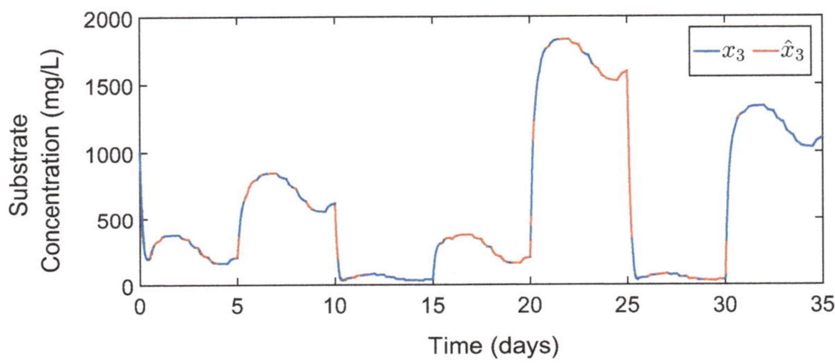

Fig. 7.2 Substrate concentration with nominal parameters and adaptive control

The initial substrate concentration is considered to be 1000 mg/L at t = 0 day. Initial concentration of anodophilic bacterial and methanogenic bacterial biomass concentration are considered to be 460 mg-xL^{-1} and 473 mg-xL^{-1} respectively. For a realistic scenario, uncertain model parameters are considered and parameter variation ranges are given in Table 7.2.

The closed-loop substrate concentration of uncertain model tracks the corresponding substrate concentration obtained from reference model with a high degree of accuracy as shown in Fig. 7.2.

The adaptive control signal (influent concentration mg/L) is shown in Fig. 7.3.

The convergence of the state error signals e_i, $i = 1, 2, 3$ to zero is shown in Fig. 7.4. Adaptive control ensures convergence of the tracking errors of the all states to zero for the given reference signal. The system with uncertain parameters follows the reference system accurately.

Adaptive controllers are inherently capable of online parameter estimation, and can ensure boundedness of the parameter estimation errors, although the algorithm does not in general ensure zero parameter estimation error at steady state conditions.

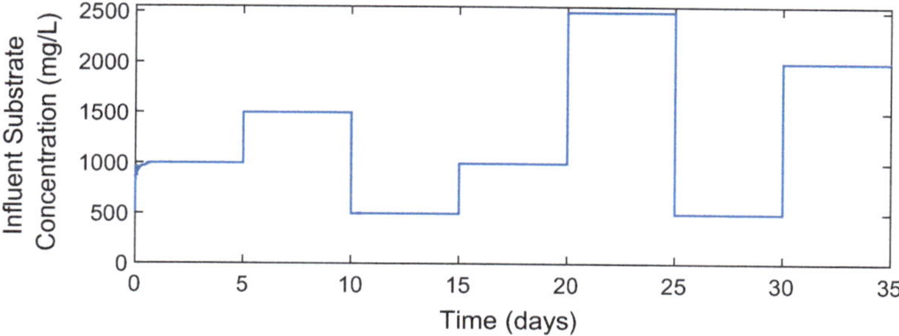

Fig. 7.3 Influent substrate concentration, $u(t)$ as determined by adaptive controller

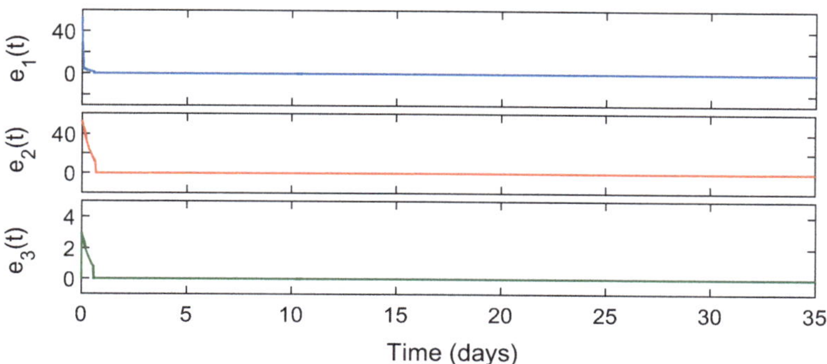

Fig. 7.4 Convergence of error signals

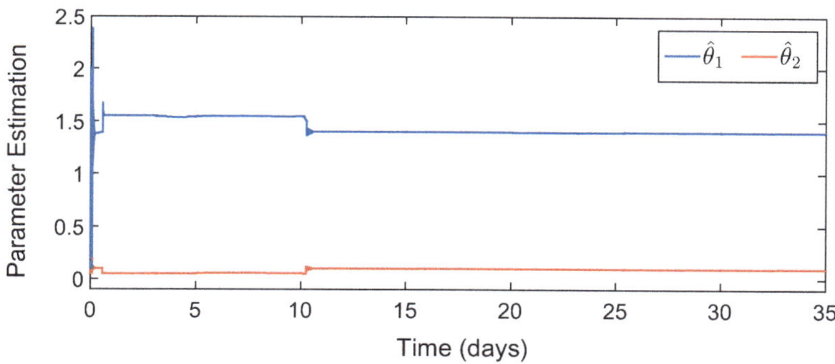

Fig. 7.5 Parameter estimation signals

Parameter bounds for this application, are provided in Table 7.2. Figure 7.5 shows parameter estimation of θ_1 and θ_2 due to adaptive action.

Figure 7.6 shows the parameter errors which approach zero at the steady state condition. It must be noted that although zero parameter error is preferred, it is not a system requirement.

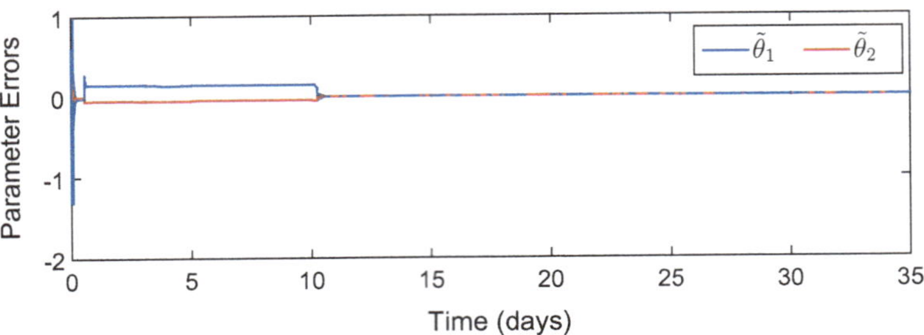

Fig. 7.6 Convergence of parameter errors

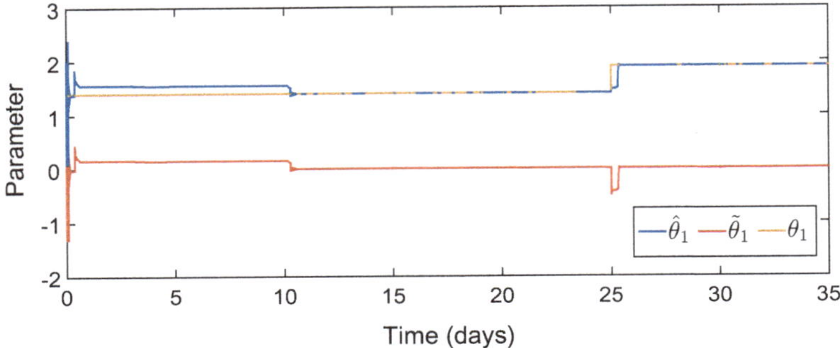

Fig. 7.7 Parameter θ_1, estimation and error with uncertainty at $t = 25$ days

Robustness of adaptive controller is ensured through a small disturbance to the uncertain parameters. To ensure controller effectiveness, we apply a small disturbance in parameter θ_1 at $t = 25$ days. The adaptive controller effectively estimates the disturbance and tries to drive the parameter error to zero after a short transient.

Figure 7.7 shows the robustness of adaptive controller against parametric uncertainty. A small disturbance is applied at $t = 25$ days and the value of θ_1 changes from 1.4 to 1.9. Adaptive laws estimate and update the uncertain parameter online and provide robustness against unmodeled dynamics and uncertain parameters.

Model based control methods developed by considering uncertainties in the MFC system parameters. Various control strategies are discussed and adaptive controller is formulated for a controllable model of the nonlinear dynamical MFC system with dual bacterial species. Performance of this control scheme under parametric uncertainty is analyzed and validated. Adaptive action provides estimations of uncertain parameters and robustness against unmodeled dynamics and uncertain parameters.

Future work can be the extension of this control scheme for multiple bacterial species with more complex MFC models. With increased model complexity, the tuning procedure of the adaptive controller gains and constants may pose a challenge that may need to be handled through proper optimization methods.

References

1. Zi-Qin, W., Sigurd, S., Ying, Z.: Exact linearization control of continuous bioreactors: a comparison of various structures. In: 2nd IEEE Conference on Control Applications, pp. 107–112 (1993)
2. Pinto, P., Srinivasan, B., Manuel, F., Tartakovsky, B.: A two-population bio-electrochemical model of a microbial fuel cell. Bioresour. Technol. **101**(14), 5256–5265 (2010)
3. Patel, R., Deb, D.: Control-oriented parametrized models for microbial fuel cells. In: 2017 6th International Conference on Computer Applications In Electrical Engineering-Recent Advances (CERA), pp. 152–157 (2017)
4. Nath, A., Deb, D., Dey, R., Das, S.: Blood glucose regulation in type 1 diabetic patients: an adaptive parametric compensation control-based approach. IET Syst. Biol. **12**(5), 219–225 (2018)
5. Deb, D., Gang, T., Burkholder, J.O., Smith, D.R.: An adaptive inverse control scheme for a synthetic jet actuator model. Am. Control Conf., 2646–2651 (2005)
6. Patel, R., Deb, D.: Parametrized control-oriented mathematical model and adaptive backstepping control of a single chamber single population microbial fuel cell. J. Power Sour. **396**, 599–605 (2018)
7. Deb, D., Gang, T., Burkholder, J.O., Smith, D.R.: Adaptive compensation control of synthetic jet actuator arrays for airfoil virtual shaping. J. Aircr. **44**(2), 616–626 (2007)
8. Lavretsky, E., Wise, K.A.: Robust and adaptive control, advance textbooks in control and signal processing (2013)

Chapter 8
Exact Linearization of Two Chamber Microbial Fuel Cell

In this chapter an exact input-output linearization control technique is developed for a DC MFC. The nonlinear mathematical model of the two compartments is discussed in Chap. 2. Exact linearization control is proposed considering dilution rate as the input variable. Control effectiveness is ascertained through simulation work.

8.1 Exact Input-Output Linearization

Nonlinear system require highly accurate control techniques and it is challenging to develop controllers for highly nonlinear systems. Linear control based on linearization around equilibrium points may be suitable for satisfactory desired performance for certain limited forms of nonlinearities. Control techniques developed primarily for linear systems, might result in bad performances in presence of higher nonlinearities and so suitable nonlinear control techniques are required. Recently, renewed interest in compensating the uncertainties of a nonlinear dynamical system, has been observed. Among such problems, an interesting one is when a change of coordinates (diffeomorphism) transforms the given nonlinear system into a linear one. There exists a series of conditions pertaining to those nonlinear systems that can be transformed via diffeomorphisms, to controllable and observable linear ones. Such a process is called exact linearization. In this method, input state or input-output is globally linearized through nonlinear coordinate transformation and state feedback.

For multi-chamber MFCs, the dynamical equations, as would be seen in subsequent Sections of this Chapter, are too complex to be suitable for nonlinear techniques like backstepping, and so other methods need to be explored. A backstepping and adaptive backstepping control technique is developed for SPSC MFC [1, 2]. Wang et al. provide a brief overview of exact input-output linearization for single input single output affine systems (nonlinear systems where input appears linearly) in [3]. The control objective is to achieve a desired steady state performance through optimization of experimental results. Let us consider a nonlinear system given by

© Springer Nature Switzerland AG 2020
R. Patel et al., *Adaptive and Intelligent Control of Microbial Fuel Cells*, Intelligent Systems Reference Library 161,
https://doi.org/10.1007/978-3-030-18068-3_8

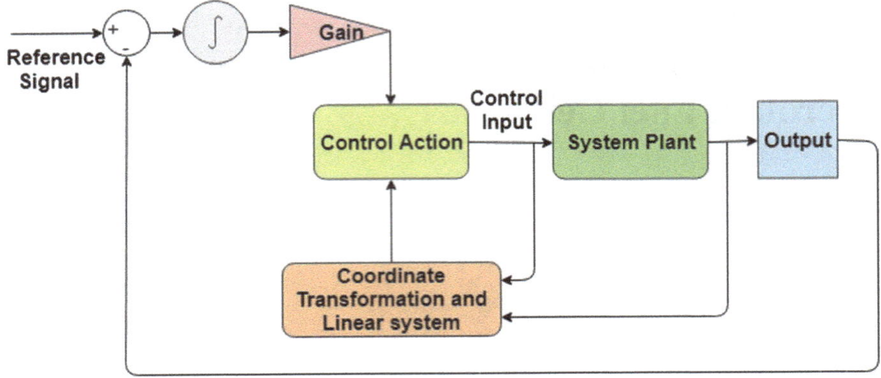

Fig. 8.1 Basic schematic diagram of exact input-output linearization control

$$\frac{dx}{dt} = a(x) + b(x)u, \quad y = c(x), \tag{8.1}$$

where x refers to the state variables, u is an input and y is the output, $a(x), b(x) \in R^n$ are smooth functions of x, and $c(x)$ is also a smooth output function.

The basic schematic diagram is shown in Fig. 8.1.

The relative degree (the difference between the poles and zeros) of the nonlinear system is r at x_0 ∀ x in a neighborhood of x_0, and these conditions are satisfied:

$$L_b L_a^i c(x) = 0, \quad \forall 0 \le i \le r - 1 \tag{8.2}$$
$$L_b L_a^{i-1} c(x) \ne 0, \tag{8.3}$$

where $L_a c$ is the Lie derivative expressed as $L_a c = \frac{\partial c}{\partial x} a(x)$. A system with finite relative degree can be transformed such that $\eta = \chi(x)$ is expressed as

$$\chi_i(x) = L_f^{i-1} h(x), \quad 1 \le i \le r$$
$$L_b \chi_i(x) = 0, \quad r + 1 \le i \le n. \tag{8.4}$$

The nonlinear state feedback control law is

$$u = \frac{1}{L_b L_a^{r-1} c} \left[v - \sum_{p=0}^{r-1} K_{p+1} L_a^p c - L_a^r c \right] \tag{8.5}$$

The desired control outcome is to be able to track the desired point y_d, and v is defined as

$$v = K_0 \int (y_d - y) d\tau. \tag{8.6}$$

The control signal renders $n - r$ states of η unobservable. These unobservable state variable must be stable for the internal system stability. Equations (8.2) and (8.3) transforms the nonlinear equations into a normal form given by

$$\frac{d\eta_i}{dt} = \eta_{i+1}, \ 1 \leq i \leq r - 1, \tag{8.7}$$

$$\frac{d\eta_r}{dt} = \alpha(\eta) + \beta(\eta)u = L_a^r c(x) + L_b L_a^{r-1} c(x) u, \tag{8.8}$$

$$\frac{d\eta_i}{dt} = \gamma(\eta), \ r + 1 \leq i \leq n \tag{8.9}$$

$$y = \eta_1. \tag{8.10}$$

Let us partition the state vector as

$$\zeta = [\eta_1, \ \dots \ \eta_r]^T \ , \quad Z = [\eta_{r+1}, \ \dots \ \eta_n]^T. \tag{8.11}$$

From (8.9) and (8.11), Z can be written as

$$\frac{dZ}{dt} = \gamma(\zeta, Z). \tag{8.12}$$

The zero dynamics are expressed as

$$\frac{dZ}{dt} = \gamma(0, Z). \tag{8.13}$$

Zero dynamics of the nonlinear system must be stable for internal stability. Zero dynamics are dependent on the control input variables and controlled output variable [3]. The existence of exact linearization is dependent on the relative degree of the system and the stability of the zero dynamics. Exact linearization control technique is applied to various nonlinear models [4–9]. Nonlinear control of bio-reactor is described using exact I/O linearization [10].

8.2 Exact Linearization Control of Anode Chamber's Dynamics

The dynamics of anode chamber is given as

$$\frac{dC_{AC}}{dt} = D_1(C_{AC}^{in} - C_{AC}) - A_m r_1, \tag{8.14}$$

$$\frac{dC_{CO_2}}{dt} = D_1(C_{CO_2}^{in} - C_{CO_2}) + 2A_m r_1, \tag{8.15}$$

$$\frac{dC_H}{dt} = D_1(C_H^{in} - C_H) + 8A_m r_1, \tag{8.16}$$

$$\frac{dX}{dt} = D_1 \frac{(X^{in} - X)}{f_x} + A_m Y r_1 - V_a K_d X, \tag{8.17}$$

where r_1 is termed as anode electrochemical reaction rate. The description of the model is given in Chap. 2. The dilution rate D_1 is used as manipulated input and substrate concentration C_{AC} is the output. Dilution rate is the direct way for controlled operation that makes substrate concentration measurement relatively reliable. Exact input-output linearization technique is applied to anode chamber dynamics. Functions $a_a(x)$ and $b_a(x)$ for the anode chamber are defined as

$$a_a(x) = -A_m r_1 \, , \ b_a(x) = (C_{AC}^{in} - C_{AC}) \tag{8.18}$$

The output and the Lie derivatives of output with respect to functions, $a_a(x)$ and $b_a(x)$ are given by

$$y_a = c_a(x) = C_{AC} \tag{8.19}$$

$$L_b c_a(x) = C_{AC}^{in} - C_{AC} \, , \ L_a c_a(x) = -\frac{A_m r_1}{V_a} \tag{8.20}$$

The control law for the anode chamber is designed from (8.5) and given as

$$D_1 = \frac{1}{C_{AC}^{in} - C_{AC}} \left[K_0 \int (y_{ad} - y_a)d\tau - K_1 y_a + \frac{A_m r_1}{V_a} \right], \tag{8.21}$$

where K_0 and K_1 are the control gains. The control law D_1 makes the input-output linear closed-loop system with below transfer function T_a expressed as

$$T_a(s) = \frac{K_0}{s^2 + K_1 s + K_0}. \tag{8.22}$$

This control technique avoids have singular point under normal operation except at the wash-out condition, $L_b y_a = C_{AC}^{in} - C_{AC} = 0$.

8.3   Exact Linearization Control of Cathode Chamber Dynamics

The dynamics of cathode chamber are given as

$$\frac{dC_{O_2}}{dt} = D_2(C_{O_2}^{in} - C_{O_2}) + A_m r_2, \tag{8.23}$$

$$\frac{dC_{OH}}{dt} = D_2(C_{OH}^{in} - C_{OH}) - 4A_m r_2, \tag{8.24}$$

$$\frac{dC_M}{dt} = D_2(C_M^{in} - C_M) + A_m N_M, \tag{8.25}$$

where r_2 is the cathode electrochemical reaction rate. The description of the model is given in Chap. 2. The dilution rate, D_2 is the control variable and oxygen concentration, C_{O_2} is the output. Exact input-output linearization technique is applied to cathode chamber dynamics. The functions $a_c(x)$ and $b_c(x)$ for the anode chamber is defined as

$$a_c(x) = -A_m\, r_2 \, , \quad b_c(x) = (C_{O_2}^{in} - C_{O_2}) \tag{8.26}$$

The output and the Lie derivatives of output with respect to $a_c(x)$ and $b_c(x)$ are

$$y_c = c_2 = C_{O_2} \tag{8.27}$$

$$L_b\, y_c = C_{O_2}^{in} - C_{O_2}, \quad L_a\, y_c = \frac{A_m\, r_2}{V_c}. \tag{8.28}$$

The control law for the input signal D_2 obtained from (8.5) is given by

$$D_2 = \frac{1}{C_{O2}^{in} - C_{O2}}\left[K_3 \int (y_{cd} - y_c)d\tau - K_4 y_c + \frac{A_m\, r_2}{V_c} \right], \tag{8.29}$$

where K_3 and K_4 are the control gains. The control law D_2 makes the input-output linear closed-loop system with below transfer function T_c expressed as

$$T_c(s) = \frac{K_3}{s^2 + K_4 s + K_3}. \tag{8.30}$$

An integrated exact linearization control of MFC is shown in Fig. 8.2.

In two chamber MFC model, the relative degree is one (ans so order of zero dynamics is also one) while considering the substrate and oxygen concentration as the controlled variable in anode and cathode chambers respectively. When dilution rate, D is the control variable, the zero dynamics are globally stable. However, if other outputs are provided, unstable zero dynamics can be controlled. Next, the closed loop simulation has been done for exact linearization control of two chamber MFC in MATLAB/Simulink. The MFC capability in voltage, current or power density depends on the chemical reaction rates of both the chambers. The reaction rate r_1 in anode chamber is dependent on the substrate concentration, biomass concentration, pH value and temperature. Assume that all other parameters in anode chamber are in controlled condition except substrate concentration.

The main control objective for developing exact linearization control of anode chamber is to accomplish the desired substrate concentration. The performance of proposed controller for anode chamber is shown in Fig. 8.3. The proposed controller

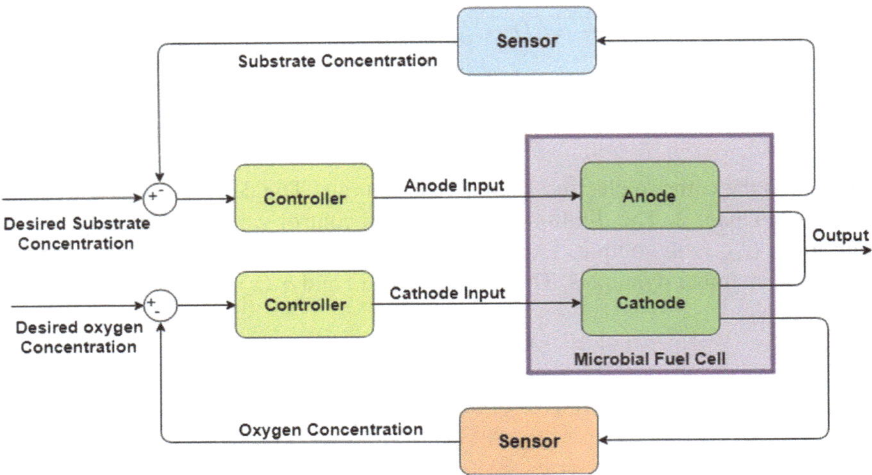

Fig. 8.2 An integrated block diagram of exact linearization control

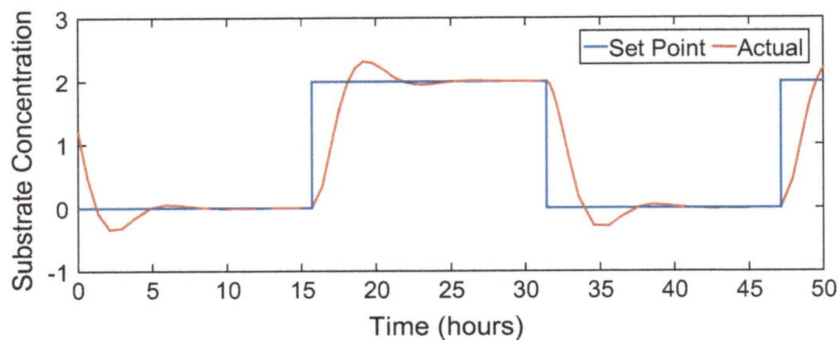

Fig. 8.3 Anode chamber controlled output (substrate concentration)

provides better tracking of substrate concentration with a set-point obtained from steady state optimization of a performance index. The control input D_1 is shown in Fig. 8.4.

The reaction rate, r_2 in cathode chamber is dependent on the oxygen concentration and temperature. Assume that all other parameters are constant in cathode chamber except oxygen concentration. Second control objective for developing exact linearization control of cathode chamber is to obtain desired oxygen concentration. The performance of proposed controller for cathode chamber is shown in Fig. 8.5. The proposed controller provides better tracking of oxygen concentration with set point. The control signal D_2 is described in Fig. 8.6. By controlling the substrate and oxygen concentration, one can get desired performance of two chamber MFC. The control gains K_1, K_2, K_3 and K_4 are 1, 1, 2 and 1 respectively.

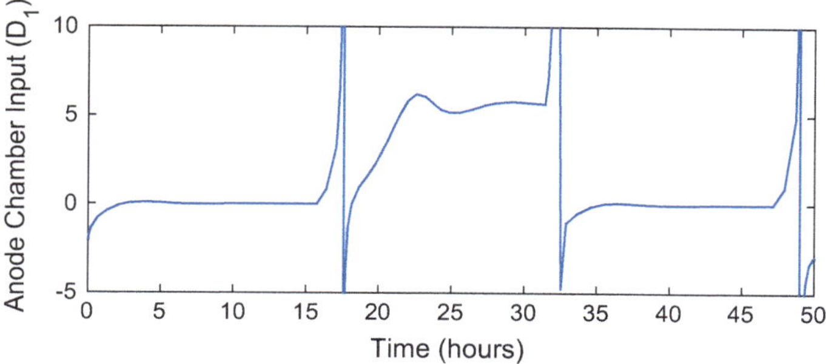

Fig. 8.4 Anode chamber control input (dilution rate D_1)

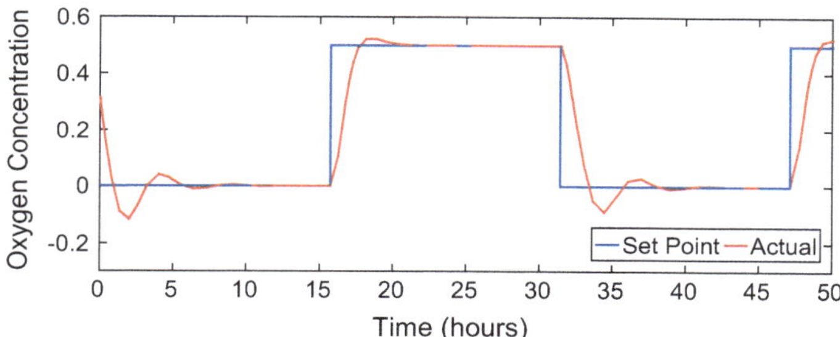

Fig. 8.5 Cathode chamber controlled output (substrate concentration)

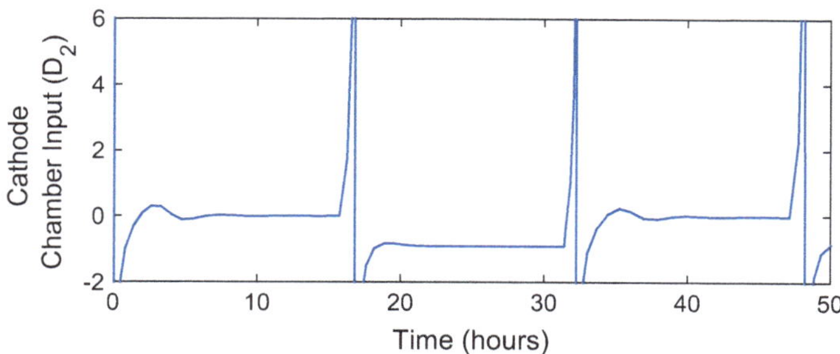

Fig. 8.6 Cathode chamber control input (dilution rate D_2)

The proposed controller provides good tracking control of substrate and oxygen concentration in anode and cathode chamber respectively. Good efficiency in terms of voltage or power density is obtained by controlling the above two concentrations.

This chapter dealt with an exact linearization control of two chamber MFC. The first part of this chapter discusses basics of the exact linearization control technique and its application on several systems. The classical linear control techniques of biological systems generally produce poor performance, even in a very small operating range. Therefore, the exact linearization control technique is widely applied to several biological systems to get better performance.

The second part of this chapter dealt with exact linearization control technique applied to two chamber MFC. The proposed control technique is separately developed for anode and cathode chamber according to their system dynamics. The performance of MFC is validated via simulation and can decided based on voltage, current and power density output depending on the substrate concentration around the anode and oxygen concentration in cathode section. The proposed controller efficiently attains and maintains desired concentration levels in respective chambers. Further, one can extend this control technique with disturbance in parameters.

References

1. Patel, R., Deb, D.: Adaptive backstepping control of single chamber microbial fuel cell. IFAC PapersOnLine **51**(1), 319–322 (2018)
2. Patel, R., Deb, D.: Parametrized control-oriented mathematical model and adaptive backstepping control strategy applied to a single chamber microbial fuel cell. J. Power Sources **396**, 599–605 (2018). Elsevier
3. Wang, Z., Skogestad, S., Zhao, Y.: Exact linearization control of continuous bioreactors: a comparison of various control structures. In: 2nd IEEE Conference on Control Applications, pp 108–112 (1993)
4. Beyer, M., Grote, W., Reinig, G.: Adaptive exact linearization control of batch polymerization reactors using a Sigma-Point Kalman Filter. J. Process. Control **18**, 663–675 (2008). https://doi.org/10.1016/j.jprocont.2007.12.002
5. Nghiep, D., Hien, N., Tan, V., Huong, N., Minh, N.: Exact linearization control for twin rotor MIMO system. SSRG Int. J. Electr. Electron. Eng. **3**, 14–19 (2016)
6. Sahibzada, A., Muhammad, J., Feng, G., Arshad, M., Bilal, K., Jacob, G., Samee K.: International Transactions on Electrical Energy Systems, vol. 26, pp. 1917–1939 (2016). https://doi.org/10.1002/etep.2185
7. Xin, Y.: Exact linearization control for a nonlinear servo system. In: 2008 Chinese Control and Decision Conference (2008). https://doi.org/10.1109/ccdc.2008.4598309
8. Mistler, V., Benallegue, A., M'Sirdi, N.: Exact linearization and noninteracting control of a 4 rotors helicopter via dynamic feedback. In: Proceedings of 10th IEEE International Workshop on Robot and Human Interactive Communication (2001). https://doi.org/10.1109/roman.2001.981968
9. Luhua, Z., Xu, C., Jiahu, G.: Simplified input-output linearizing and decoupling control of wind turbine driven doubly-fed induction generators. In: 2009 IEEE 6th International Power Electronics and Motion Control Conference (2009). https://doi.org/10.1109/ipemc.2009.5157462
10. Pröll, T., Karim, N.: Nonlinear control of a bioreactor model using exact and I/O linearization. Int. J. Control **60**, 499–519 (1994). https://doi.org/10.1080/00207179408921478

Chapter 9
Microbial Fuel Cell Laboratory Setup

In this chapter, MFC development in laboratory scale using cow manure as a waste, is described. Details of the materials used for the experiment is provided with their size and volume. The behavior of MFC is studied as per the experimental data of inputs and output. The transfer functions of anode and cathode chamber are obtained through a system identification approach.

9.1 Materials

This section provides the information related to materials used for the MFC experiment. Selection of materials of anode and cathode electrodes, and membrane are critical in determining the overall functioning of the MFCs. The basic properties of aforementioned materials have already been given in Chap. 1. The details of materials and chemicals are given in Table 9.1.

MFCs have different types of configuration such as two chamber, single chamber, cube or cylindrical model, H-type model, up-flow model etc.. Extensive research has taken place to increase the performance and try to scale up with proper design configuration [1]. In this setup, a two chamber MFC with an up-flow configuration is developed. The main objective of MFC is to generate bioelectricity from waste. Dairy wastewater, farm or domestic wastewater, cow manure, hospital wastewater etc. can be used as a waste [2]. We have used cow manure of local dairy farm (Ahmedabad, Gujarat, India) as a waste.

Ultra-filtration membrane holder (diameter 47 mm and volume 250 ml) has two compartments. A membrane filter paper separates the compartments. Anode compartment consists of pre-treated cow manure as a waste, a mixture of substrate and anode media, and stainless steel mesh with dimension of 4 cm × 4 cm as an electrode. Cathode compartment consists activated carbon cloth of dimension 4 cm × 4 cm as an electrode and cathode media shown in Fig. 9.1.

© Springer Nature Switzerland AG 2020

R. Patel et al., *Adaptive and Intelligent Control of Microbial Fuel Cells*, Intelligent Systems Reference Library 161,

https://doi.org/10.1007/978-3-030-18068-3_9

Table 9.1 Materials list for the MFC setup

Material	Specification/ Application/ Nomenclature
Membrane filter holder	47 mm × 250 ml
Membrane filter paper	47 mm
Stainless steel mesh sheet	Anode electrode
Activated carbon Cloth	Cathode electrode
Single core copper wire	Wire gauge depends on design requirement
Digital multimeter	For measurement of voltage and current
Distilled water	—
Potassium dihydrogen phosphate	KH_2PO_4
Di-potassium hydrogen phosphate	K_2HPO_4
Potassium nitrate	KNO_3
Magnesium sulphate	$MgSO_4$
Calcium chloride	$CaCl_2$
Sodium chloride	Nacl
Sodium bicarbonate	$NaHCO_3$
D-Glucose	$C_6H_{12}O_6$

Fig. 9.1 Stainless steel mesh and activated carbon cloth

The cathode chamber is mounted above the anode chamber like a sediment type MFC configuration. Ultra-filtration membrane of diameter 47 mm is used as a separator between anode and cathode chamber. D-glucose and KNO_3 are used as feeding stock or substrate for energy of bacteria in anode and cathode respectively. The chemicals along with their concentration for making anode and cathode media are given in Table 9.2.

Pre-treatment process is required to remove methanogens and develop a suitable environment for bacterias available in cow manure. As per standard procedure, 60

Table 9.2 Anode and cathode media

Ingredients	Concentration (g/l)
KH_2PO_4	4.4
K_2HPO_4	3.4
Nacl	0.5
$MgSO_4$	0.2
$CaCl_2$	0.014
$NaHCO_3$ (Only for cathode), KNO_3	1
Trace metal solution (Only for anode)	1 ml/l
D-Glucose (Feeding Stock for anode)	100
KNO_3 (Feeding Stock for cathode)	100

gram cow manure is required for making 500 ml cow manure slurry [3]. Cow manure slurry contains 20–25% wet weight in 50 mM KH_2PO_4 at pH value 4.5. This cow manure slurry is kept for an incubation period of 10–15 h under aerobic condition and room temperature on shaker. The purpose of this process is to suppress methanogens in the cow manure. However, aerobic sparging period may not completely kill the methanogens but slow down the methane emission rate. After the sparging period, the slurry is diluted with 50 mM K_2HPO_4 at pH value 7.0.

9.2   Procedure and Operation

The anode compartment is filled with 125 ml pre-treated cow manure slurry and 125 ml anode media. D-Glucose is used as a food for the bacterias present in cow manure slurry. At the outset, make a feeding stock solution of D-Glucose depending on the number of MFC setups and volume of each chamber. The concentration of D-Glucose is varied depending on the goal of the experiments. For example, to make 60 ml feeding stock solution of D-Glucose, 6 g D-Glucose and 60 ml distilled water is required. The complete MFC setup is shown in Fig. 9.2.

The following formula provides the relationship between concentration and feeding stock required in a MFC setup:

$$100 \times X = C \times V, \tag{9.1}$$

where X is the feeding stock in *ml* for single setup, C refers to the concentration required for feeding stock in anode compartment and V refers to the volume of combination of the cow manure slurry and anode media. It is required to have 5 g/l concentration of D-Glucose for 250 ml volume, and 12.5 ml feeding stock is taken from the 60 ml feeding stock solution. One can choose concentration based on one's experimental requirements.

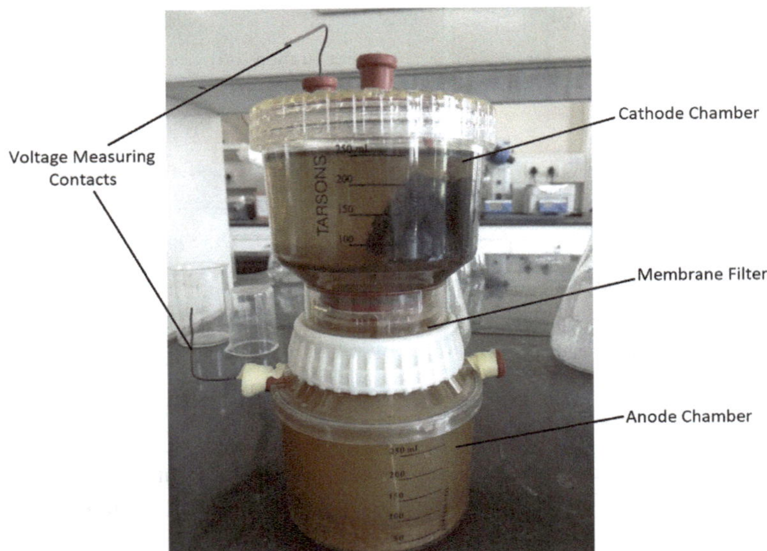

Fig. 9.2 MFC laboratory setup

The cathode chamber is fitted with an ultra-filtration membrane separating the anode chamber such that there is no leakage. There is conductivity between anode and cathode chamber through membrane. Cathode chamber is filled with 240 ml cathode media, 10 ml cow manure slurry and feeding stock of KNO_3. Preparation of feeding stock procedure and concentration formula are same as the anode chamber. An overview of experimental process of MFC is shown in Fig. 9.3.

If one has access to other types of bacteria, one can add those instead of cow manure slurry. Anaerobic condition is needed throughout the process, and so air cathode is used. The operation is mainly based on the oxidation and reduction reactions in the two compartments. The first reaction is oxidation in the anode compartment wherein microorganisms consume substrate for energy and water, thereby generating hydrogen ions, electrons, and carbon dioxide as a byproduct. The generated ions are pass through the membrane and are attracted to the cathode and reduction takes place during which, hydrogen ions combine with electrons. The electrons are transferred through the anode surface and copper wire connected between anode and cathode.

For the experimental setups, the size and materials required are different based on applications. For sensing applications, typically benthic mud or top soil samples are used [4]. However, we used cow manure collected from local dairy farm. Performance enhancement of MFCs is obtained through different materials of anode and cathode electrodes, PEM or waste source. We have also ensured the cost effectiveness along with performance enhancement. Most of the applications of MFCs widely depend on their internal resistance and output current density. The objective of effective cost of MFCs is fulfilled by decreasing the internal resistance and increasing the

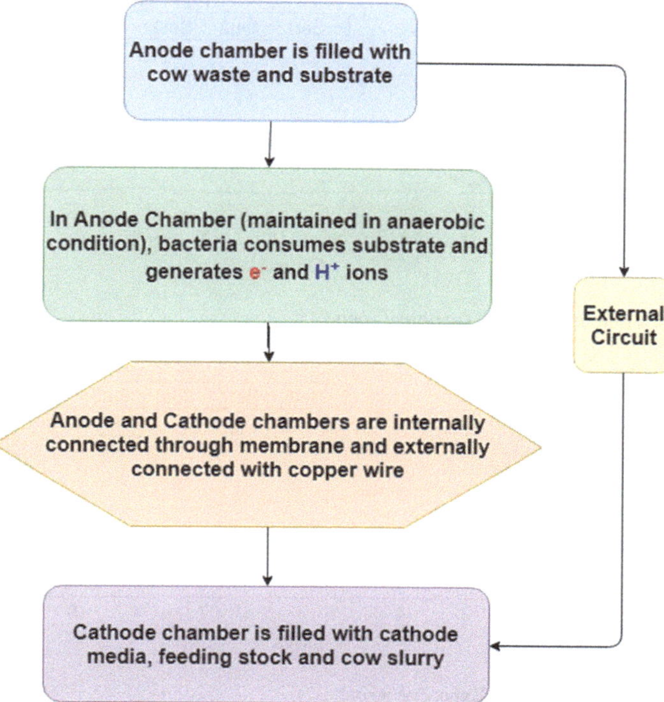

Fig. 9.3 Overview of experimental process of MFC

output current density. After successful experimental studies, scaling of MFCs is the next step which is more challenging because of resistance of electrolyte, anode, cathode and membrane. There are certain solutions available that can help reduce the resistance level such as electrolyte resistance by using lower conductive wastewater. Scaling of MFCs depend on factors like types of applications and improvements in material desired and the cost of materials [5].

9.3 Experimental Results

Five MFC setups are developed in each round of experiments and a total of three rounds are conducted. Fifteen sets of readings are reported and the voltage generated from MFCs are shown in Figs. 9.4, 9.5 and 9.6.

The average voltage across a total of 10 days from different setups are given in Table 9.3. These datasets are used to develop a transfer function model of the MFC using system identification method.

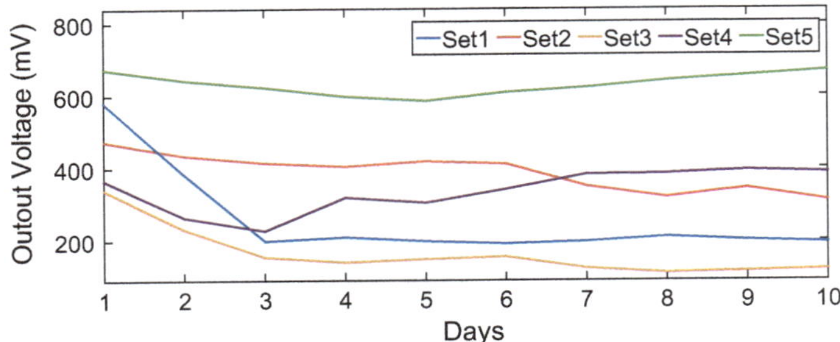

Fig. 9.4 MFCs output voltage (Experiment round 1)

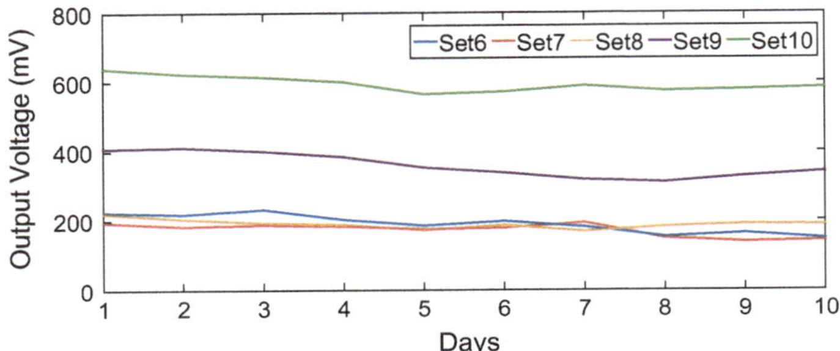

Fig. 9.5 MFCs output voltage (Experiment round 2)

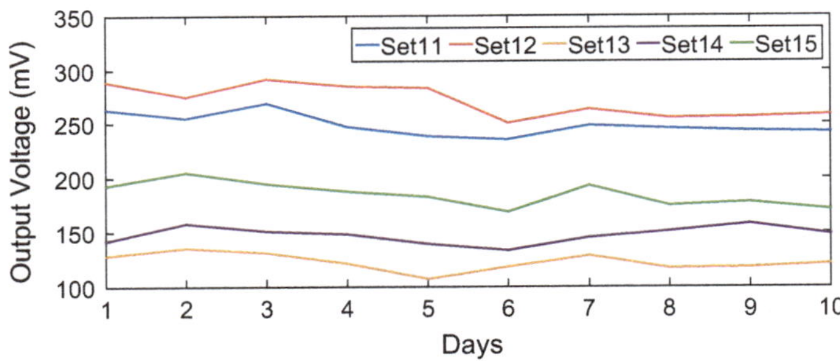

Fig. 9.6 MFCs output voltage (Experiment round 3)

9.4 System Identification

Suitable mathematical models of nonlinear systems are important to decipher system behavior as well as for the effective control of system operations and to estimate the output as compared to the reference input. A mathematical model encompasses complex operations into simple equations to characterize the impact of operational

Table 9.3 Output voltages with different Inputs from MFC setups

Setup number	Anode chamber input (Glucose Concentration)	Cathode chamber input (KNO$_3$ Concentration)	MFC voltage (mV)
Set 1	1	3	256.8
Set 2	2	4	388.4
Set 3	3	5	163.8
Set 4	4	1	357.7
Set 5	5	2	631.5
Set 6	1	2	190.8
Set 7	2	5	173.4
Set 8	3	1	188.7
Set 9	4	4	359.1
Set 10	5	3	593.7
Set 11	3.5	2.5	248.5
Set 12	1.5	3.5	270.5
Set 13	0.5	1.5	122.5
Set 14	4.5	0.5	146.6
Set 15	2.5	4.5	185.2

and design parametric components of the equation on the output performance, while identifying the relationship between system parameters, input and output variables.

System modeling and identification is done as per the type and knowledge of the system [6]. Modern control methods like model predictive control, adaptive control, sliding mode control etc. are model based. For industrial applications, modeling and system identification are important aspects of control. There are two ways to obtain models for real-time plants:

1. Using first principles of chemistry, physics etc. called white box modeling;
2. Estimation of models using experimental data called black box modeling. Sometimes combination of these two are also used [7].

System identification (SI) consists of the following steps: (i) optimal experimental design, (ii) data collection related to inputs and outputs, (iii) model characterization, (iv) model estimation or identification, and (v) model verification. The first and second steps are most important to obtain system dynamics. It involves accurate designing of inputs and experimental conditions otherwise it leads to wrong system dynamics. The second step is to define system characteristics like number of zeros, system order or type etc., and to choose a model structure like state space, polynomial, transfer function, process models, spectral models, correlation models, and different types of auto-regressive models. Model estimation involves estimating parameters or systems with best fits according to inputs and outputs. Model verification provides

Fig. 9.7 System
identification process flow

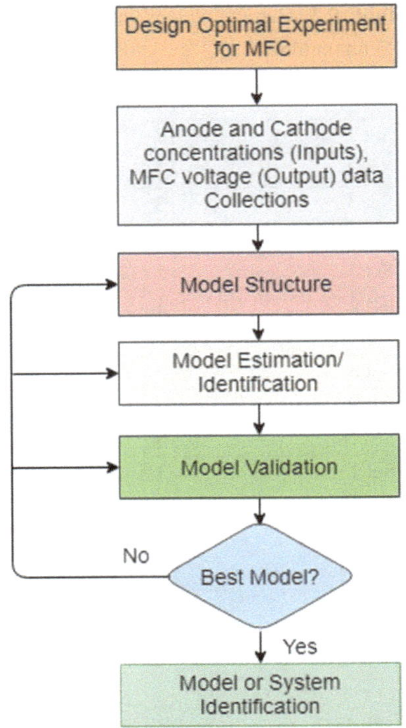

the time or frequency domain responses of the identified system with the best fits
[8]. Basic flow of operation in this process is shown in Fig. 9.7.

System identification toolbox provides a mathematical representation of dynamical system in terms of transfer functions. Model estimation depends on characteristics such as number of zeros and system order. Estimation of transfer function and its parameters using system identification is discussed in different research articles [9–12]. MFC is a multiple input single output (MISO) system. The inputs (combinations of anode and cathode concentration) and output (MFC voltage) experimental data was collected as given in Table 9.3. Combination of different number of poles and zeros is taken to obtain best fit transfer function. The combination of poles and zeros with estimation accuracy is given in Table 9.4.

Next, transfer function model with accuracy 90.4% is considered as the most appropriate. Two chamber MFC has 4th and 3rd order of anode and cathode dynamics given in Chap. 3, which supports our estimated transfer function model. One can develop control scheme by using these transfer function models. Anode chamber transfer function (T_a) and cathode chamber transfer function (T_c) are given by

Table 9.4 Estimation of MFC model

Number of poles [Anode, Cathode]	Number of zeros [Anode, Cathode]	Estimation	Label accuracy (%)
[4, 3]	[1, 1]	63.17	tf1
[3, 2]	[1, 1]	40.43	tf2
[3, 2]	[0, 0]	38.53	tf3
[4, 3]	[0, 0]	90.4	tf4
[4, 3]	[2, 2]	53.39	tf5
[4, 3]	[3, 2]	56.72	tf6

$$T_a(s) = \frac{4.16 \times 10^4}{s^4 + 11.44s^3 + 60.71s^2 + 180.9s + 312.1} \tag{9.2}$$

$$T_c(s) = \frac{-94.97}{s^3 + 2.619s^2 + 4.09s + 4.819}. \tag{9.3}$$

These transfer functions are then used for developing suitable control strategies for effective performance of MFCs.

This chapter dealt with system identification approach to find out transfer function models of the two compartments of MFCs. The first part of the chapter discussed the laboratory setups of MFCs. Materials and chemicals required for the experimental setup is given in Table 9.1 with their specifications. The detailed procedure and operation is explained with an appropriate example. After several steps and procedures, input-output reading of 15 sets are collected. These readings are used to find the transfer function model through system identification technique.

System identification tools provide various models from their input output datasets. Combination of a different number of poles and zeros is taken to get the highest accuracy and best fit transfer function model of MFC. The most accurate transfer function model is selected for development of suitable control techniques.

References

1. Nandy, A., Kundu, P.: Configurations of microbial fuel cells. Progress Recent Trends Microb. Fuel Cells, 25–45 (2018)
2. Aelterman, P., Rabaey, K., Clauwaert, P., Verstraete, W.: Microbial fuel cells for wastewater treatment. Water Sci. Technol. **54**, 9–15 (2006)
3. Vijay, A., Vaishnava, M., Chhabra, M.: Microbial fuel cell assisted nitrate nitrogen removal using cow manure and soil. Environ. Sci. Pollut. Res. **23**, 7744–7756 (2016)
4. Jessica, L.: An experimental study of microbial fuel cells for electricity generating: performance characterization and capacity improvement. J. Sustain. Bioenerg. Syst. **3**, 171–178 (2013)
5. Sleutels, T.H.J.A., Ter Heijne, A., Buisman, C.J.N., Hamelers, H.V.M.: Bioelectrochemical systems: an outlook for practical applications. ChemSusChem **5**(6), 1012–1019 (2012)
6. Wang, L., Zhao, W.: System identification: new paradigms, challenges, and opportunities. Acta Autom. Sinica **39**, 933–942 (2013)

7. Schoukens, J., Vandersteen, G., Barbé, K., Pintelon, R.: Nonparametric preprocessing in system identification: a powerful tool. Eur. J. Control. **15**, 260–274 (2009)
8. Saengphet, W., Tantrairatn, S., Thumtae, C., Srisertpol, J.: Implementation of system identification and flight control system for UAV. In: 3rd International Conference on Control, Automation and Robotics, pp. 678–683 (2013)
9. Ozdemir, A., Gumussoy, S.: Transfer function estimation in system identification toolbox via vector fitting. IFAC PapersOnLine **50**(1), 6232–6237 (2017)
10. Deb, A., Roychoudhury, S., Sarkar, G.: System identification: parameter estimation of transfer function. In: Analysis and Identification of Time-Invariant Systems, Time-Varying Systems, and Multi-Delay Systems using Orthogonal Hybrid Functions. Studies in Systems, Decision and Control, vol. 46. Springer, Cham (2016)
11. Schijndel, A.W.M.: The Use of system identification tools in MatLab for transfer functions. In: 3rd Annual Meeting of Climate for Culture Project (CfC), EU-FP7-Project no.: 226873, Visby, Sweden, pp. 1–34 (2011)
12. Fruk, M., Vujisić, G., Špoljarić, T.: Parameter identification of transfer functions using MATLAB. In: 36th International Convention on Information and Communication Technology, Electronics and Microelectronics, pp. 697–702 (2013)

Chapter 10
Model Reference Adaptive Control of Microbial Fuel Cells

In this Chapter, two kinds of MRAC techniques of MFC are presented. Basics of MRAC scheme is already given in Chap. 5. The transfer function models of anode and cathode chambers are discussed in previous Chapter. The first technique is MRAC using MIT rule and the second one is Lyapunov based MRAC technique. The performance of both the developed control schemes is validated through appropriate simulation work.

10.1 Model Reference Adaptive Control Using MIT Rule

MRAC scheme contains three major parts namely- reference system, actual system with controller and parameter adjustment block. It contain two different control loops. The inner loop is the feedback control loop which comprises of the actual system and the controller, whereas the outer loop consists of an adjustment mechanism for parameters. The block diagram of this scheme using MIT rule is shown in Fig. 10.1.

The parameters are adjusted in such a manner that desired performance and error between actual system and the reference system eventually goes to zero. Actual system has known structure, whereas the system parameters may not be known. The reference model is used to obtain a desired performance of the adaptive controller in reference or desired reference input. The gradient method (MIT rule) and Lyapunov stability method are used for MRAC design. The gradient method was conceived at by Massachusetts Institute of Technology, MIT and widely used for autopilot aircraft [1]. This method is applied to various plants such as DC motor, power plant super heater, magnetic levitation etc. [2–5]. In this method, a cost function $J(\theta)$ of adjustable parameter, is defined as

$$J(\theta) = \frac{1}{2}e^2, \tag{10.1}$$

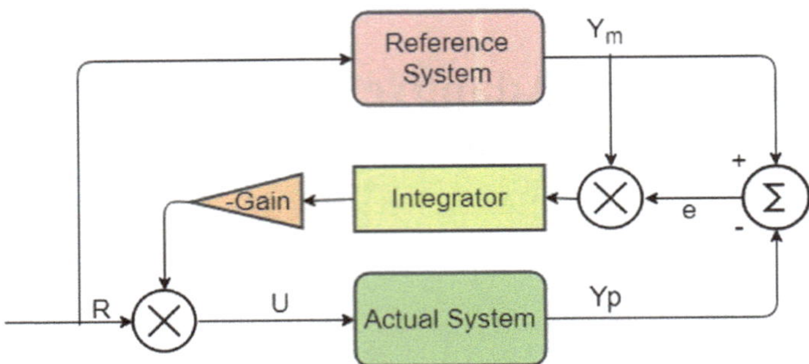

Fig. 10.1 Block diagram of MRAC using MIT rule

where e refers to the error between output of the actual system and the reference system. The main objective is to regulate the parameter θ such that the cost function has a minimum value, and so we define

$$\frac{d\theta}{dt} = -\gamma \frac{\partial J}{\partial \theta} = -\gamma e \frac{\partial e}{\partial \theta}, \tag{10.2}$$

where $\frac{\partial e}{\partial \theta}$ is defined as the sensitivity derivative indicating the changes of error with respect to parameter θ and γ is the adaptive gain of control action.

The transfer function of the actual system is given as

$$Y(s) = K_p G_p(s) U(s), \tag{10.3}$$

where $Y(s)$, $G_p(s)$, and $U(s)$ are the actual system output, actual system transfer function, and input signal of the system respectively, and K_p represents the unknown parameter of the system. Similarly, the reference system is defined as

$$Y_m(s) = K_m G_m(s) R(s), \tag{10.4}$$

where $Y_m(S)$, $G_m(s)$, and $R(s)$ are the reference system output, reference system transfer function, and reference input signal of the system respectively, and K_m represents the known parameter of the reference system. The error signal $e(t)$ is represented as

$$e(t) = Y(t) - Y_m(t), \tag{10.5}$$

where $Y(t)$ and $Y_m(t)$ refer to the output of actual and reference system respectively. The error function in s-domain is defined as

$$E(s) = Y(s) - Y_m(s). \tag{10.6}$$

From (10.3) and (10.4), and using a control law, $U = \theta R$, we obtain

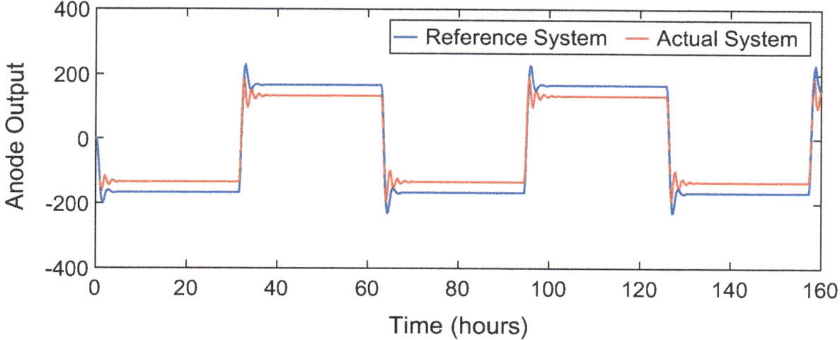

Fig. 10.2 Anode chamber performance without control action

$$E(s) = K_p G_p(s)\theta R(s) - K_m G_m(s) R(s). \tag{10.7}$$

The partial derivative of $E(s)$ is given as

$$\frac{\partial E(s)}{\partial \theta} = K_p G_p(s) R(s) = \frac{K_p}{K_m} Y_m(s) \tag{10.8}$$

Substitute (10.8) in (10.2), we can get

$$\frac{d\theta}{dt} = -\gamma \frac{K_p}{K_m} Y_m(s) = -\gamma_1 e Y_m \tag{10.9}$$

where γ_1 represents the adaptive control gain, $\gamma \frac{K_p}{K_m}$.

10.2 Simulation Results

In this section, we set out to validate the performance of MRAC technique using MIT rule for the two chamber MFC. The transfer functions of anode and cathode chamber are given in the previous chapter. The adaptive controller gains for anode and cathode chamber (γ_1) are 0.0005 and 0.000015 respectively. It helps to improve the transient behavior and closed-loop performance.

The control goal is to obtain and maintain a expected steady state condition of the original system following the reference system which has desired properties. The performance of anode and cathode chamber output without MRAC using MIT rule is shown in Fig. 10.2 and Fig. 10.3 respectively.

It is noticed that the performance of actual systems do not follow the reference systems and there is a finite error between them. Now, the control actions have to minimize error in the performance. The tracking performance of anode and cathode chamber outputs is given in Fig. 10.4 and Fig. 10.5 respectively. It is fact that from

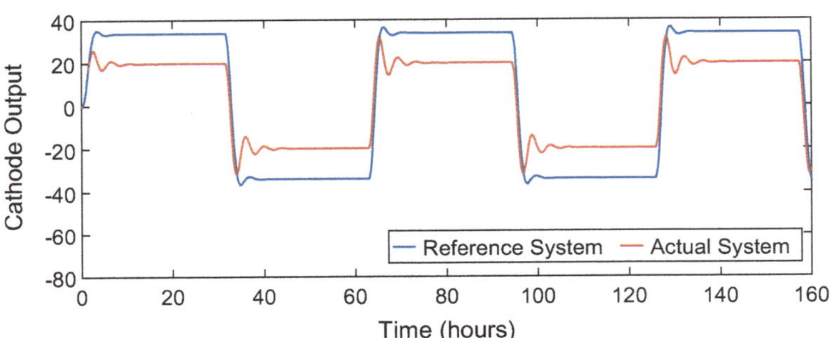

Fig. 10.3 Cathode chamber performance without control action

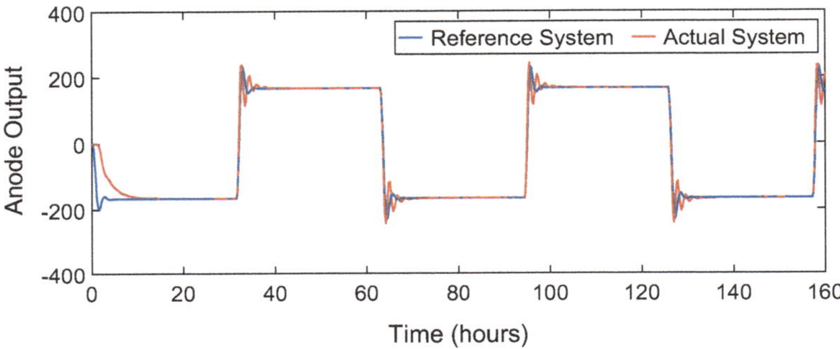

Fig. 10.4 Anode chamber performance with MRAC using MIT rule

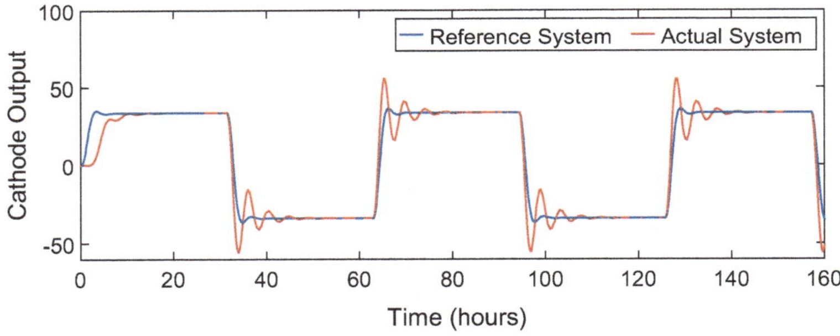

Fig. 10.5 Cathode chamber performance with MRAC using MIT rule

results the performance of actual systems of both the chambers follows the respective reference systems with a high degree of accuracy.

The control signals generated by the adaptive control mechanism as the input of actual system and the reference input signal are shown Fig. 10.6. The convergence of the state error signal e to zero is shown in Fig. 10.7. The MRAC control mechanism causes the asymptotic tracking error with respect to the given reference signal to go

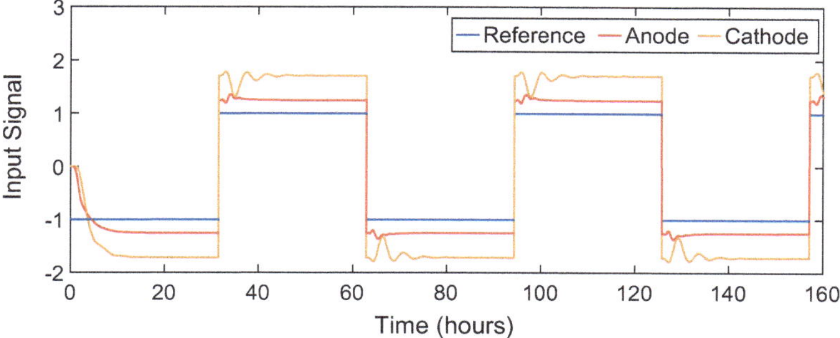

Fig. 10.6 Control Signal from adaptive control

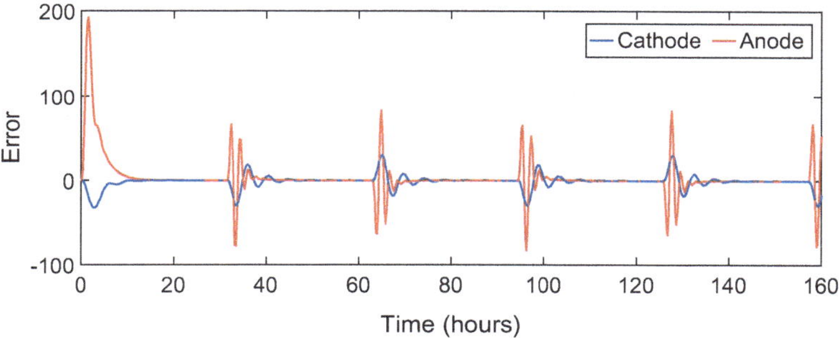

Fig. 10.7 Convergence of error signal

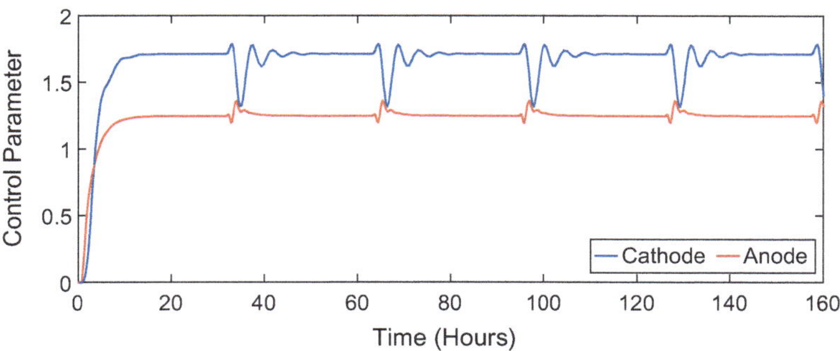

Fig. 10.8 Variation in control parameters of both the chambers

to zero. The fluctuation in error signal is due to the transients available in every cycle. The effect of adaptive gain on time response graph for MIT rule is shown Fig. 10.8. The increment in adaptive gains shows the improvement in system performances but the range of adaptive gains which provide desired performance is limited.

The stated goals of the simulation studies are to validate the developed controller's ability in estimating the control parameters for system stability and error minimization. Choice of the reference input signal, $r(t)$ plays vital role to achieve the control objective which can be decided from a detailed knowledge of microbial fuel cells. After some initial transients, the output signal is followed by the reference signal. In this method, proper selection of adaptive gains is very important to obtain better results. For studied model of MFC, MRAC with MIT rule gives satisfactory performance and the plant follows the chosen reference model precisely.

10.3   Model Reference Adaptive Control

An adaptive control technique is developed as a stabilization or tracking control, wherein the adaptation takes place on the tracking error between the plant and reference model output. In MRAC, a good knowledge about behavior and performance of the reference model is required, such that the desired input-output properties of the closed-loop system is achieved. By designing a proper reference model, one can develop effective adaptive control mechanism. Typically, a reference model is a linear time-invariant model but it can be also a nonlinear reference model. The basic configuration of MRAC scheme applied to anode and cathode chambers of MFC is shown in Fig. 10.9.

Such a control mechanism contains three major subsystems namely, actual plant, reference plant, and adaptive mechanism for parameter estimation. The objective of adaptive control methodology is to maintain the tracking error as small as possible by proper adaptation mechanism in presence of uncertain parameters. Adaptive laws are ordinary differential equations that allow adjustment of adaptive parameters so as to maintain the tracking error minimum. Stability of the adaptive control mechanism is mathematically analyzed by Lyapunov stability theory. The number of adaptive laws depend on the number of uncertain or unknown parameters to be estimated on-line. Model reference adaptive control is applied to various plants [6–9]. The comparison of MRAC with MIT rule and Lyapunov method is done in [10].

Consider a linear system given as

$$\dot{X}_j = A_j X_j + B_j U_j, \tag{10.10}$$

where $X_j \in \mathbb{R}^n$, $A_j \in \mathbb{R}^{n \times n}$, $B_j \in \mathbb{R}^{n \times m}$, and $U_j \in \mathbb{R}^n$ refer to state vector, system matrix, control matrix, and control input, subscript j refers to a (for anode) and c (for cathode). The state feedback control input is defined as

$$U_j = -K_j^T X_j, \tag{10.11}$$

where K_j is the gain vector. Assume that the matrices A_j and B_j are such that the system is controllable. The main control objective is that the actual system effectively tracks the reference system. The reference system which has all desired properties,

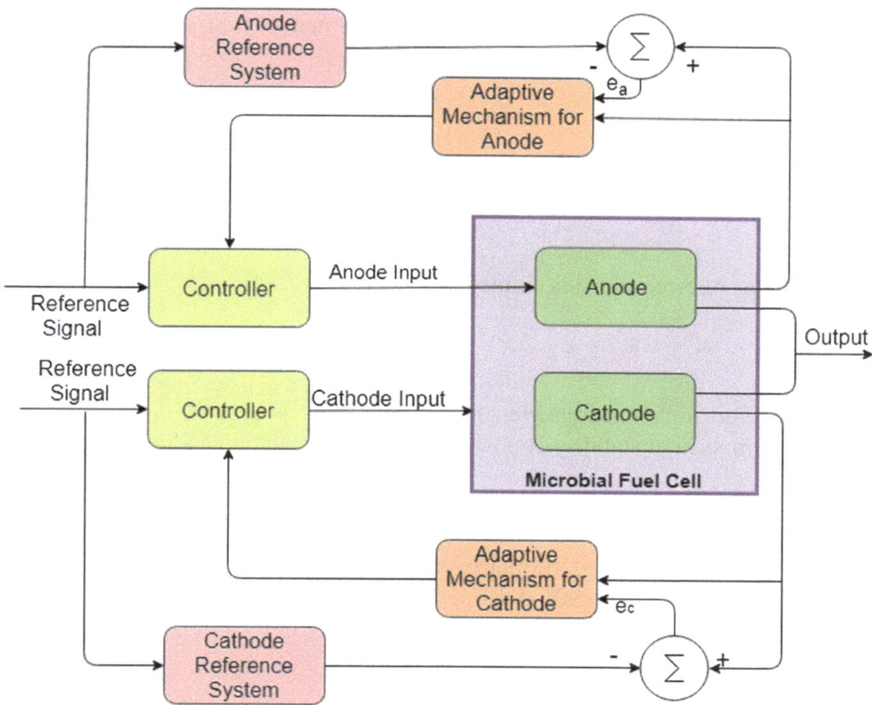

Fig. 10.9 Basic configuration of MRAC scheme for anode and cathode chambers

is designed as

$$\dot{X}_{rj} = A_{rj} X_{rj} + B_{rj} R_j, \tag{10.12}$$

where R_j refers to the reference input signal. The tracking error is defined as

$$e_j = X_j - X_{rj}. \tag{10.13}$$

To ensure effective tracking, all the signals must be uniformly bounded and the tracking error should be

$$\lim_{t \to \infty} ||e_j|| = 0, \tag{10.14}$$

for both 'a' (anode) and 'c' (cathode). Substituting, (10.11) in (10.10), the system is given as

$$\dot{X}_j = (A_j - B_j K_j^T) X_j. \tag{10.15}$$

From (10.12) and (10.15), we can obtain that

$$A_{rj} = A_j - B_j K_j^T. \tag{10.16}$$

If the vector K_j is not known, the control law is defined as

$$U_j = -\hat{K}_j^T X_j, \tag{10.17}$$

where $\hat{K}_j^T \in \mathbb{R}^{m \times n}$ refers to the estimation of gain K_j^T. Substituting, (10.17) into (10.10), we obtain

$$\dot{X}_j = (A_j - B_j \hat{K}_j^T) X_j. \tag{10.18}$$

The derivative of error signal is defined as

$$\dot{e}_j = \dot{X}_j - \dot{X}_{rj} = A_{rj} e_j - B_j \tilde{K}_j^T X_j, \tag{10.19}$$

where $\tilde{K}_j^T$ refers to the parameter error.

To confirm system stability, a positive definite Lyapunov candidate function is chosen

$$V_j = e_j^T P_j e_j + tr(\tilde{K}_j^T \Gamma_j^{-1} \tilde{K}_j) \tag{10.20}$$

where tr refers to the trace of the matrix which is defined as the addition of the main diagonal values. Note that, only diagonal elements are effective for the stability of the system. The matrix $P_j \in \mathbb{R}^{n \times n}$ satisfies $P_j = P_j^T > 0$,

$$P_j A_{rj} + A_{rj}^T P_j = -Q_j, \tag{10.21}$$

and $Q_j \in \mathbb{R}^{n \times n}$ refers to constant matrix with $Q_j = Q_j^T > 0$. The derivative of V_j is given as

$$\dot{V}_j = \dot{e}_j^T P_j e_j + e_j^T P_j \dot{e}_j + 2tr(\tilde{K}_j^T \Gamma_j^{-1} \dot{\hat{K}}_j). \tag{10.22}$$

Substituting, (10.13), (10.19), and (10.21) into (10.22), we obtain

$$\dot{V}_j = -e_j^T P_j e_j + 2tr\left(\tilde{K}_j^T [\Gamma_j^{-1} \dot{\hat{K}}_j - X_j e_j^T P_j B_j]\right). \tag{10.23}$$

The adaptive law is chosen in such a manner that the system stability is ensured through $\dot{V} < 0$. An appropriate adaptive law in this case is given as

$$\dot{\hat{K}}_j = \Gamma_j X_j e_j^T P_j B_j, \tag{10.24}$$

and the $\dot{V}_j$ becomes

$$\dot{V} = -e_j^T Q_j e_j < 0. \tag{10.25}$$

Additionally, since, $e_j(t)$, and $\hat{K}_j(t)$ are uniformly bounded, and $\lim_{t \to \infty} e_j(t) = 0$.

It can be seen that all the signals are bounded thereby guaranteed system stability and convergence of errors to zero for both the anode and the cathode. One of the major benefit of the Lyapunov stability theory is that it requires little computing power and ensures stability of the uncertain system. Lyapunov technique assures tracking error convergence rather than parameter convergence.

10.4 Performance Evaluation and Simulation Results

Simulation work is done to ensure the efficacy of the developed control technique for transfer function models of the two chamber MFC. The matrices A_a and B_a for anode chamber obtained from transfer function $T_a(s)$ are

$$A_a = \begin{bmatrix} -11.44 & -60.71 & -180.9 & -312.1 \\ 1 & 0 & 0 & 0 \\ 0 & 1 & 0 & 0 \\ 0 & 0 & 1 & 0 \end{bmatrix}, \quad B_a = \begin{bmatrix} 1 \\ 0 \\ 0 \\ 0 \end{bmatrix}.$$

The gain vector, K_a and adaptive gain, Γ_a is determined as

$$K_a = \begin{bmatrix} 4.05 & 2.8 & -6.1 & -3.2 \end{bmatrix}, \quad \Gamma_a = \begin{bmatrix} 0.05 & 0 & 0 & 0 \\ 0 & 1 & 0 & 0 \\ 0 & 0 & 1 & 0 \\ 0 & 0 & 0 & 1 \end{bmatrix}.$$

The chosen reference system is

$$A_{ra} = \begin{bmatrix} -15.44 & -80.71 & -180.9 & -200.1 \\ 1 & 0 & 0 & 0 \\ 0 & 1 & 0 & 0 \\ 0 & 0 & 1 & 0 \end{bmatrix}.$$

The matrix P_a is selected in such a manner that ensures $P_a A_{ra} + A_{ra}^T P_a = -Q_a$:

$$P_a = \begin{bmatrix} 10 & 4 & 2.5 & 1 \\ 10 & 4 & 2.5 & 1 \\ 10 & 4 & 1 & 2.5 \\ 10 & 5.20 & 1 & 1 \end{bmatrix},$$

$$Q_a = 1.0e + 03 \times \begin{bmatrix} 0.0808 & 0.7753 & 1.7984 & 1.9884 \\ 0.1348 & 0.7984 & 1.8041 & 1.9994 \\ 0.1348 & 0.7999 & 1.8026 & 1.9994 \\ 0.1338 & 0.7999 & 1.8064 & 1.9971 \end{bmatrix}.$$

The matrices A_c and B_c for cathode chamber obtained from transfer function $T_c(s)$ given in previous chapter, are

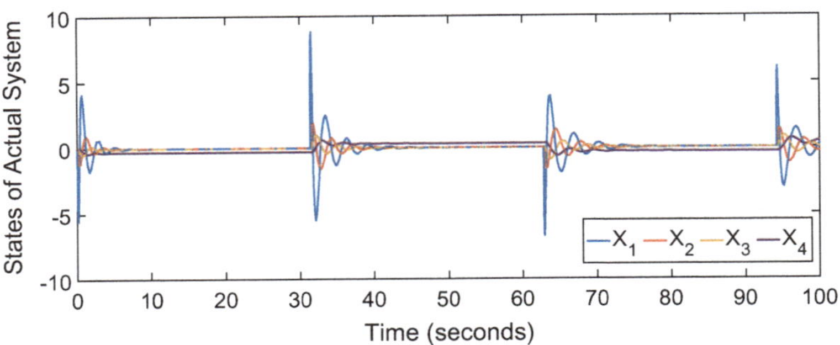

Fig. 10.10 Performance of anode states

$$
A_c = \begin{bmatrix} -2.69 & -4.09 & -4.81 \\ 1 & 0 & 0 \\ 0 & 1 & 0 \end{bmatrix}, \quad B_c = \begin{bmatrix} 1 \\ 0 \\ 0 \end{bmatrix}.
$$

The gain vector K_c, adaptive gain Γ_c and the reference system is determined as

$$
K_c = \begin{bmatrix} 0.33 & 0.21 & 0.05 \end{bmatrix}, \quad \Gamma_c = \begin{bmatrix} 0.05 & 0 & 0 \\ 0 & 1 & 0 \\ 0 & 0 & 1 \end{bmatrix}, \quad A_{rc} = \begin{bmatrix} -3.61 & -4.09 & -4.81 \\ 1 & 0 & 0 \\ 0 & 1 & 0 \end{bmatrix}.
$$

The matrix P_c is chosen so as to ensure the condition $P_c A_{rc} + A_{rc}^T P_c = -Q_c$:

$$
P_c = \begin{bmatrix} 25 & 10 & 0.5 \\ 20 & 2 & 4.5 \\ 10 & 4.5 & 2 \end{bmatrix}, \quad Q_c = \begin{bmatrix} 19.1598 & 63.7557 & 106.3559 \\ 31.4448 & 61.7259 & 95.4213 \\ 0.5418 & 35.7852 & 41.0917 \end{bmatrix}.
$$

The control objective is the tracking of actual systems following the desired reference systems. The performance of anode and cathode states are shown in Fig. 10.10 and Fig. 10.11 respectively.

The MRAC control scheme ensures the asymptotic tracking errors with respect to the given reference signal, go to zero. The convergence of the anode and cathode errors to zero are presented in Fig. 10.12 and Fig. 10.13 respectively. Transient fluctuations are observable in each cycles. Error signal is seen to go to zero, that is the actual system effectively tracks the reference system.

The comparison of anode and cathode reference and generated input signal through adaptive control are given in Fig. 10.14 and Fig. 10.15 respectively.

Choice of reference signal is important to achieve control objective which can be decided from a thorough knowledge of behavior of MFCs. The control signals of anode and cathode are followed by the respective reference signals after some initial transients and desired performance of MFC is obtained. Additionally, adaptive

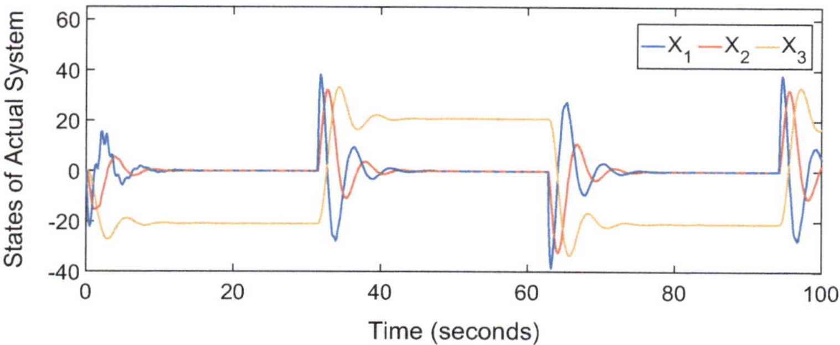

Fig. 10.11 Performance of cathode states

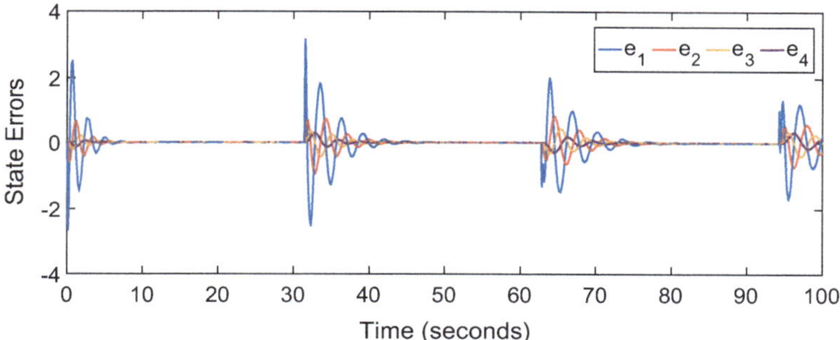

Fig. 10.12 Convergence of anode state errors

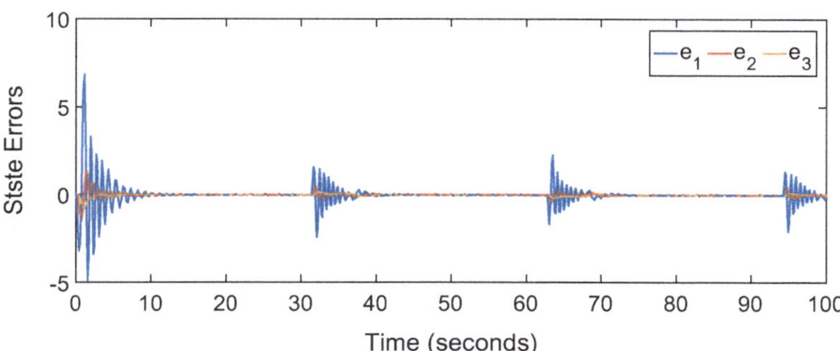

Fig. 10.13 Convergence of cathode state errors

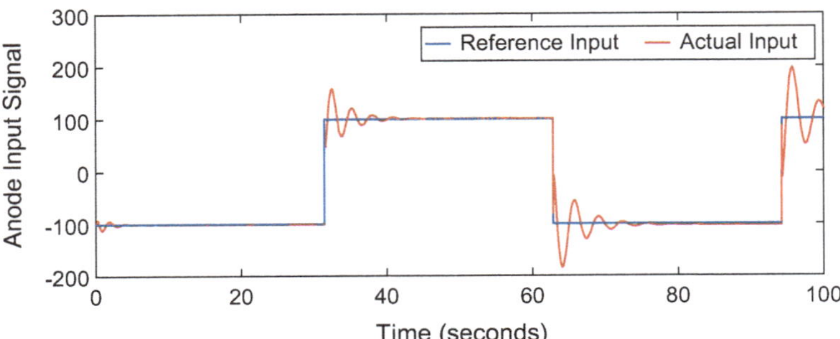

Fig. 10.14 Anode input signals

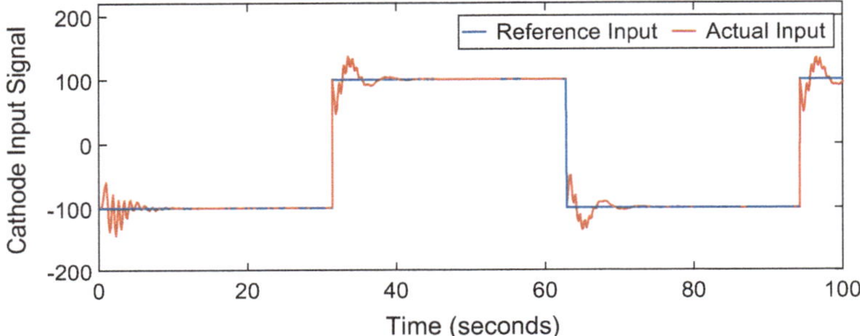

Fig. 10.15 Cathode input signals

control mechanism provides robustness against parametric uncertainty. Adaptive control laws ensure the output convergence of the uncertain system.

This chapter dealt with two different kind of MRAC techniques used for the transfer functions of MFCs which are obtained via system identification approach. The first part of the chapter discusses the development of model reference adaptive control technique with MIT rule for anode and cathode chamber. Simulation work evaluates the performance of the developed controller vis-a-vis effectively tracking of the reference system and ensures the convergence of the state errors to zero. Adaptive gains of the controller and reference input signals are important factors in system performance, so that proper selection is necessary to get better performance. The second part of the chapter deals with the MRAC technique with Lyapunov stability analysis. The performance of the developed controller is justified through appropriate simulation work. The adaptive mechanism provides robustness against parametric uncertainty. After initial transients, the controller provides smooth tracking performance and ensures convergence of state errors to zero.

The primary limitations for commercialization of MFCs are the scale up of the process and uninterrupted functioning while maintaining adequate microbial concentration. For the purpose of operation for a sustained period of time, it is critical to understand the system dynamics through further exhaustive experiments and analyzing the data thus obtained. However, performing such experiments is time consuming and expensive, and so the other approach of precise modeling the system to understand the dynamics, needs further work.

References

1. Swathi, M., Ramesh, P.: Modeling and analysis of model reference adaptive control by using MIT and modified MIT rule for speed control of DC motor. In: 7th International Advance Computing Conference, pp. 482–486 (2017)
2. Priyank, J., Nigam, M.: Design of a model reference adaptive controller using modified mit rule for a second order system. Adv. Electron. Electr. Eng. **3**(4), 477–485 (2013)
3. Zdenek, M., Martin, P., Stepan, O.: Simulation of MIT rule-based adaptive controller of a power plant superheater. Front. Comput. Educ. (2012)
4. Coman, A., Axente, C., Boscoianu, M.: The simulation of the adaptive systems using the MIT rule. In: 10th WSEAS International Conference on Mathematical Methods and Computational Techniques In Electrical Engineering, pp. 301–305 (2008)
5. Singh, B., Kumar, V.: A real time application of model reference adaptive PID controller for magnetic levitation system. In: 2015 IEEE Power, Communication and Information Technology Conference (PCITC) (2015)
6. Kavuran, G., Alagoz, B., Ates, A., Yeroglu, C.: Implementation of model reference adaptive controller with fractional order adjustment rules for coaxial rotor control test system. Balkan J. Electr. Comput. Eng. **4**(2), 84–88 (2016)
7. Sapiee, M., Abdullah, F., Noordin, A., Jahari, A.: PI Controller design using model reference adaptive control approaches for a chemical process. In: 2008 Student Conference on Research and Development (2008)
8. Erik, S., Ming, L., Weijia, T., Fuchen, C., Jie, F., Cagdas, O.: Adapting to flexibility: model reference adaptive control of soft bending actuators. IEEE Robot. Autom. Lett. (2017)
9. Coman, S., Boldisor, C.: Model reference adaptive control for a DC electric drive. Bull. Transilvania Univ. Braşov Ser. I Eng. Sci. **6**(55), 33–38 (2013)
10. Tariba, N., Bouknadel, A., Haddou, A., Ikken, N., Omari, H., Omari, H.E.l.: Comparative study of adaptive controller using MIT rules and Lyapunov method for MPPT standalone PV systems. In: AIP Conference Proceedings, pp. 04008-1–04008-07 (2017)